Adiabatic Invaria

Atmospheri

Adiabatic Invariants in Large-scale Atmospheric Dynamics

Michael V. Kurgansky

Department of Atmospheric and Oceanic Physics,
University of Concepcion, Concepcion, Chile

and

A. M. Obukhov Institute of Atmospheric Physics,
Moscow, Russia

CRC Press
Taylor & Francis Group
Boca Raton London New York

CRC Press is an imprint of the
Taylor & Francis Group, an **informa** business
A TAYLOR & FRANCIS BOOK

CRC Press
Taylor & Francis Group
6000 Broken Sound Parkway NW, Suite 300
Boca Raton, FL 33487-2742

First issued in paperback 2020

© 2002 by Taylor & Francis Group, LLC
CRC Press is an imprint of Taylor & Francis Group, an Informa business

No claim to original U.S. Government works

ISBN-13: 978-0-367-45475-3 (pbk)
ISBN-13: 978-0-415-28415-8 (hbk)

**Visit the Taylor & Francis Web site at
http://www.taylorandfrancis.com**

**and the CRC Press Web site at
http://www.crcpress.com**

Publisher's Note
This book has been prepared from camera-ready copy supplied by the author

Every effort has been made to ensure that the advice and information in this book is true and accurate at the time of going to press. However, neither the publisher nor the authors can accept any legal responsibility or liability for any errors or omissions that may be made. In the case of drug administration, any medical procedure or the use of technical equipment mentioned within this book, you are strongly advised to consult the manufacturer's guidelines.

British Library Cataloguing in Publication Data
A catalogue record for this book is available from the British Library

Library of Congress Cataloging in Publication Data
A catalog record for this book has been requested

CONTENTS

FOREWORD TO THE ENGLISH EDITION   vii

PREFACE   ix

INTRODUCTION   1

CHAPTER 1. Equations of Motion and Conservation Laws   9
    1.1. Equations of compressible fluid dynamics   9
    1.2. Adiabatic approximation. Potential vorticity theorem   18
    1.3. Vorticity charge (potential vorticity substance)   23
    1.4. Helicity   26
    1.5. Energy and entropy of the atmosphere   31
    1.6. Atmospheric angular momentum   43

CHAPTER 2. Reduced Equations of Atmospheric Dynamics   51
    2.1. Two-dimensional atmospheric models
        ('shallow-water' approximation)   52
    2.2. Adjustment of fluid dynamic fields   62
    2.3. Filtered (quasi-geostrophic) equations   72
    2.4. Rossby waves   78
    2.5. Continuous (spontaneous) wave emission   82

CHAPTER 3. Hydrodynamic Instability of Conservative Motions   89
    3.1. Barotropic instability   89
    3.2. Baroclinic instability   97
    3.3. Linear analysis (Eady model)   106
    3.4. Two-layer Phillips' model   110
    3.5. Vertical stability of atmospheric motions.
        Richardson's criterion   115

CHAPTER 4. Isentropic Analysis of Large-Scale Processes   123
    4.1. Isentropic coordinates   123
    4.2. Theorem on potential vorticity   128
    4.3. Precise formulation of the available potential
        energy concept   130
    4.4. Diabatic transformation of potential vorticity   136

4.5. General properties of air adiabatic motion   141
    1. Gauge invariance in potential vorticity definition   141
    2. Topological invariants   143
    3. Account for compressibility of motion on isentropic
       surfaces   147
    4. Invariant flux tubes and their deformation   149
4.6. Isentropic potential vorticity maps and invariant
    flux tubes   154
4.7. Distribution of adiabatic invariants in the atmosphere   167
4.8. The concept of atmospheric vorticity charge and its
    applications   169

CHAPTER 5. Dissipative Processes in the Atmosphere   **187**
5.1. Surface boundary layer   187
5.2. Ekman boundary layer   193
5.3. Ekman friction mechanism   198
5.4. Turbulent Ekman layer   202

REFERENCES   207

INDEX   219

FOREWORD TO THE ENGLISH EDITION

When preparing the English edition of the book, the author aimed to give the reader a concise but advanced introduction to large-scale extratropical atmospheric dynamics from the viewpoint of a representative of the Russian scientific school on dynamic meteorology and fluid dynamics. An increasingly broadening stream of publications on the subject made it impossible to give in this edition a survey of all interesting and relevant papers published in the 1990s. The author referred only to a few of them, as the most important and representative from his point of view. In the recent years, the most drastic development seems to have happened in the field of atmospheric climate variability modeling and investigations of strato-spheric dynamics and chemistry. In the context of the latter problem, isen-tropic Ertel's potential vorticity maps, plotted every day and distributed by the European Centre for Medium Range Weather Forecasts (ECMWF) based in Reading, UK, have become nowadays a common tool even for those researchers who are far from being experts in the field of atmospheric dynamics. Recently, relatively less attention has been paid to the use of potential vorticity in the diagnosis of large-scale tropospheric processes and especially in the climate theory. This gives the author an excuse to leave Chapter 4 nearly unchanged, and only to give in Section 4.8 some recent data on the statistics of potential vorticity in the atmosphere, obtained by the author and his colleagues. To provide a 'trampoline' for the reader and not begin with formulae, a short, seven-page, Introduction has been also written for the English edition. All other minor pieces of additional material and certain comments are given in the footnotes.

The time to dot the i's and cross the t's in the book fell on the author's stay at the University of Concepción, Department of Atmospheric and Oceanic Physics, and he cordially thanks his Chilean colleagues for collab-oration. Certainly, the same words should be addressed to the author's colleagues and old friends at the A. M. Obukhov Institute of Atmospheric Physics in Moscow.

August 2001

PREFACE

The aim of the book is to give an introduction to the dynamics of large-scale atmospheric circulation systems with spatial dimensions from one to several thousand kilometres. Eddies of this horizontal extent are permanently observed in the atmosphere and determine the weather and climate on the Earth. Currently, a satisfactory enough theory on extratropical large-scale atmospheric motions exists. It became possible because of the clear physical principles laid into its foundation. They combine the account of closeness between actual and geostrophic wind (and the corresponding idea of the quasi-geostrophic essence of large-scale atmospheric motions) with the use of fundamental fluid dynamic constants of motion – of potential vorticity and energy, first and foremost. Account of these principles, including the summary on the fundamentals of the theory of atmospheric motion constants and their application in meteorology, is the task which the author had addressed himself. An extremely difficult problem, in the author's opinion, of large-scale equatorial atmosphere dynamics has been left 'overboard'. Here, besides other things, it appears necessary to take into account water vapor phase transitions. Moreover, Ertel's theorem on the conservation of potential vorticity is, generally speaking, not valid for moist air, and the quasi-geostrophic theory cannot be applied in this case. However, this problem is a constituent of the problem of general atmospheric circulation. To create a consistent theory in this field remains a task for the future.

The theme of the monograph is broader than it follows from its title. It might be considered as an introduction to the dynamics of quasi-two-dimensional fluid flows. This term implies, first of all, nearly horizontal displacements of fluid parcels in a vertically stratified compressible fluid. From the standpoint of fluid dynamics the atmosphere is just such a system. We do not touch upon the subject of essential three-dimensionality, which is specific for meso- and small-scale atmospheric motions. This issue is extraordinarily complex, because in the framework of a three-dimensional problem one has actually unlimited vortex tube stretching; besides, the vorticity field could have a very complex topological structure. It is the topic of numerous current investigations in the field of fluid dynamics and turbulence theory.

An essential restriction made in the book and consistent with the quasi-two-dimensionality concept (when strong air updrafts and downdrafts are suppressed) is to treat atmospheric air as effectively dry. By all means, quasi-horizontal layered structure of the atmosphere can be locally violated and this is what is regularly observed in nature. Here, we mean tropopause folding, fronts, intense cyclones, etc., where account for air humidity is necessary. Among problems considered in the book, the concept of helicity stands somewhat aside. It should appear most fruitful when essentially three-dimensional motions are considered, but this is beyond the subject of the book. Nevertheless, even for quasi-two-dimensional motions the helicity concept has a nontrivial meaning being a measure of flow 'non-self-similarity' in different fluid layers. To appreciate fully the benefit of the helicity concept for the atmospheric dynamics is the matter of the future but even today it is clear that to create a consistent theory of intense atmospheric vortices, such as hurricanes, tornadoes, twisters, etc. is impossible without using it.

In its present form, the book has largely originated from the lectures on atmospheric dynamics given by the author during a number of years to graduate students of the Chair of Atmospheric Physics in the Physical Department at Moscow University. It so happened that two distinguished experts in the field of dynamic meteorology, first A. F. Dubuc and then A. M. Obukhov, held this Chair in the past years. Together with L. A. Diky who had also been associated with the Chair, they established both research and academic traditions, which the author of these lines is trying to follow. However, in the process of writing the book, this circumstance led to certain difficulties. The author's ebullience with his personal research problems permanently conflicted with the sense of measure usual for a reader. The author's 'research egoism' evidently dominated when writing Chapters 1, 3 and, particularly, 4 (see Table of Contents). Chapters 2 and 5 are nearer to a textbook. To the highest degree, it could be attributed to Chapter 5 which gives the fundamentals of the planetary boundary layer theory. This chapter has been written in order to make the book more complete and it presents the well-known, classical results in a rather compressed form. This has determined a more free style of citing references there, permissible when writing textbooks. An exclusion is the final section of Chapter 5 which presents some recent results of the author's colleagues at the Institute of Atmospheric Physics, Russian Academy of Sciences, on the theory of turbulent Ekman layer.

The author is grateful to his many colleagues at the Institute of Atmospheric Physics for their moral support, patience and condescension expressed when the manuscript was in work, which took much more time than initially expected. First of all, kind words should be addressed personally to F. V. Dolzhanskii who read the first draft of the manuscript and gave a lot of valuable comments. The author is very much obliged to his teacher, A. M. Obukhov, and this book is dedicated to his memory.

Recollections of scientific discussions with him and the very image of this remarkable scientist and outstanding individual threw light on many pages of the manuscript.

September 1992

INTRODUCTION

Atmospheric phenomena range from tiny, highly transient turbulent eddies up to planetary-scale quasi-permanent atmospheric structures with a horizontal size comparable to the Earth's radius. Fluid-dynamically, the finest-size atmospheric vortical motions are characterized by *Kolmogorov's microscale of turbulence*, of the order of one millimeter, named after A. N. Kolmogorov (1907–1987).[1] This is the minimum dimension of vortices, which can survive against air molecular viscosity, provided the mechanical power supply, due to the non-linear cascade of energy from grosser vortices, is a prescribed quantity specified by the requirements of the overall energy balance in the atmosphere. The minimum lifetime of vortices is of the order of seconds. The gravest atmospheric features are *jet streams*, on a spatial scale of thousands of kilometers and a timescale of months. These are great 'rivers of air', which flow from west to east and circumnavigate the entire Earth.

Nowadays, a classification of atmospheric phenomena with regard to scale limits, suggested in 'A rational subdivision of scales for atmospheric processes' (Orlanski, 1975), is widely accepted. Here, the characteristic horizontal distance is divided into the large, or macro scale, including the phenomena with horizontal extension greater than 2,000 kilometers; the small, local or microscale, including the features smaller than 2 kilometers; and the middle, or mesoscale, which includes all atmospheric phenomena with scales between the macroscale and the microscale. Classification according to the timescales is more ambiguous and less frequently used. In this context, the macroscale usually refers to atmospheric features with periods of several days and longer, and the microscale corresponds to phenomena with a duration from seconds to minutes and the mesoscale corresponds to lifetime from half an hour to one week. In Orlanski (1975), this threefold division of space-scale is further continued by subdividing the

---

[1] See also Frisch (1998), where a modern account of turbulence is given and Kolmogorov's 1941 theory of 'fully developed turbulence' is clearly explained, along with the discussion of interesting historical aspects concerning its creation and further development.

large-scale into the macroα-scale, with lengths greater than 10,000 kilometers, and the macroβ-scale, with space scale ranging between 2,000 and 10,000 kilometers. In this book, we shall call macroβ-scale the synoptic or large-scale, properly speaking, and call macroα-scale the planetary scale. The meso- and microscales are further divided into α, β and γ subscales, where the alphabetic order corresponds to a gradual decrease in size. An interested reader requiring more details is referred to the original paper by Orlanski (1975), special monographs by Pielke (1984) and Atkinson (1989) and the thematic article 'Scales' by R. Avissar in Encyclopedia of Climate and Weather (1996).

The supply of energy from the Sun to the terrestrial atmosphere occurs at the gravest space-scale of 10,000 kilometers – the distance between the equator and poles – with a net excess of heat in the tropics and a net deficit poleward of 40° latitude. This uneven distribution of heat source and sink is a primary cause of the *general atmospheric circulation*, see also a classical monograph 'The Nature and the Theory of the General Circulation of the Atmosphere' by Edward Lorenz (1967).

As the atmospheric scale height, of the order of 10 kilometers, is small compared to the Earth's radius (≈6400 kilometers), only a vertical component of the doubled Earth's rotation angular velocity vector, called the *Coriolis parameter*, plays a dynamically significant role in the general atmospheric circulation. As a consequence, one observes two distinct circulation regimes in the terrestrial atmosphere. The first of these, namely the *Hadley regime*, is characterized by direct meridional thermal circulation with updrafts over the equator and descending branches over the subtropics. A large amount of energy fed into the tropical atmosphere comes from latent heat release, accompanying micro- and mesoscale deep (penetrative) moist convection, associated with cloud clusters. In the tropics, the Coriolis parameter is small and, moreover, it strictly vanishes at the equator. Consequently, the background Earth's rotation is not very important dynamically. In this region, large-scale patterns are smooth, rather immobile and persistent, and the greatest variability comes from the mesoα-scale phenomena, such as *hurricanes* and *typhoons*.

The large-scale dynamics of the tropical atmosphere has its long-standing unresolved problems, though many powerful researchers worked in this field and very good theories have been proposed. For example, a full satisfactory theoretical explanation of a somewhat mysterious phenomenon known as *Quasi-Biennial Oscillation*, which is a nearly rhythmic alternation of zonal wind direction in the lower equatorial stratosphere of about 28 months periodicity, still remains a challenge to experts in atmospheric dynamics.

A quite different circulation regime, the so called *Rossby regime*, is observed in the extratropics and is associated with well-pronounced large-scale features, namely *cyclones* and *anticyclones*, which are permanently observed in the atmosphere over mid and high latitudes.

In this book we focus on the dynamics of large-scale atmospheric flows in the extratropical atmosphere. The reason why these motions can be treated in isolation from mesoscale phenomena, at least approximately, deserves explanation. In brief, the macroscale range has its own main feed of energy and also its principal energy sink, with no mesoscale motions being involved. This is due to the instability of the time-mean zonal circulation of the atmosphere and conversion of the stored potential energy into kinetic energy of the large-scale eddies. Another somewhat intrinsically related reason is as follows. Due to rapid diurnal rotation of the Earth, apparent in the atmospheric dynamics in mid and high latitudes, meteorological fields depend on altitude in a quite different manner as compared to latitude and longitude. Namely, the changes in meteorological parameters at different altitudes are strongly coupled via the *thermal wind* relation, and a great deal of similarity between atmospheric patterns at different horizontal levels is observed. As a result, the large-scale motions are not truly three-dimensional but nearly two-dimensional or, as it is sometimes said, $2\frac{1}{2}$-dimensional. According to *Kelvin's circulation theorem,* the kinematic constraint of the approximate conservation of area of a material fluid element at each horizontal level restricts the vertical stretching (compressing) of vortex tubes with the horizontal cross-section of a large diameter. Thus, a non-linear cascade of energy toward smaller scales is effectively suppressed, energy is trapped in the macroscale and, moreover, it is likely to cascade toward the gravest space scales, inversely. This can also be interpreted in terms of a 'spectral gap' in the power spectra of meteorological variables, which separates between macro- and mesoscales (see Atkinson, 1989). The principal sink of kinetic energy of large-scale flows is maintained by viscous friction between the atmosphere and the underlying solid Earth and occurs within the *planetary boundary layer*, at the expense of a direct transfer of energy from large-scale circulation systems to small-scale turbulence.

A great success of dynamic meteorology in the 20th century was a rational explanation as to why the synoptic-scale atmospheric motions observed, notably cyclones and anticyclones, have a dominant space-scale of a few thousand kilometers, that is, one order of magnitude smaller than the scale at which the energy from the Sun is fed. This was a principal result of the baroclinic instability theory pioneered by J. G. Charney (1917–1981) in 1947 and independently E. T. Eady (1915–1966) in 1949. The major prediction was that the space-scale of most rapidly growing baroclinic disturbances is bounded from below by the characteristic length

$$\frac{Brunt - Vaisala\ (buoyancy)\ frequency \times atmospheric\ scale\ height}{Coriolis\ parameter},$$

which is known as the Rossby internal, or baroclinic, radius of deformation. Typically, the atmospheric scale height is about 8 kilometers. For a geographic latitude of 45°, the magnitude of the Coriolis parameter is very close to 10^{-4} sec^{-1}. The *Brunt–Vaisala frequency* characterizes the vertical oscillations of a blob of air affected by the gravitational restoring force due to stable stratification of the atmosphere, and the usual magnitude of this frequency is 10^{-2} sec^{-1}. So, roughly speaking, the *Rossby radius* is slightly less than 1,000 kilometers.

In the extratropics, large-scale atmospheric flows are nearly geostrophic, i.e. are permanently observed in the vicinity of an exact geostrophic balanced state when the deflective Coriolis force equilibrates the pressure gradient force. Typically, deviations from geostrophicity are about 10%. This implies the smallness of the ratio between the *relative vorticity* and the Coriolis parameter, which is essentially the *planetary vorticity*. The latter clearly reflects evidence that in large-scale atmospheric vortices, namely cyclones and anticyclones, air parcels revolve relative to the earth 10 times slower than the solid Earth itself rotates in the space. Thus, from the viewpoint of large-scale atmospheric dynamics, the Earth is quite a rapidly rotating planet.

To a very good approximation, macroscale flows in the free atmosphere behave as if air were an ideal fluid, without dissipation. However, regardless of how small molecular viscosity is, it must be accounted for in the close vicinity of the Earth's surface, where the vector of wind velocity vanishes due to non-slip boundary conditions. In the bulk of a planetary boundary layer, with a typical height of 1 kilometer, the surface friction reduces wind speed and, hence, the magnitude of the Coriolis force acting on fluid parcels. The relative shallowness of the planetary boundary layer guarantees that the horizontal pressure gradient force remains unchanged and keeps the value characteristic of the free atmosphere. The resulting imbalance between the pressure gradient and the Coriolis force induces a cross-isobar flow component from high toward low pressure. The continuity of mass requires a vertical mass flow at the top of the planetary boundary layer: downward in regions of high pressure and upward in regions of low pressure. Through the work done by these cross-isobar motions against the frictional forces, the free atmosphere is constantly transferring kinetic energy to the planetary boundary layer, where it is ultimately converted into heat.

Summarizing, we conclude that in the extratropics the large-scale air motions are thermally driven, nearly geostrophic flows of stable stratified and mildly viscous atmosphere, when the presence of a thin planetary boundary layer does not destroy the quasi-geostrophicity in the free atmosphere but maintains a principal energy sink.

Note how all of these conditions are violated in the equatorial atmosphere: (i) quasi-geostrophicity fails due to low values of the Coriolis parameter; (ii) stable stratification is regularly violated in the regions of moist penetrative convection; (iii) moist convective cells transport momentum in a vertical direction so efficiently that the entire equatorial troposphere may be well-considered as the planetary boundary layer.

When studying the large-scale atmospheric motions, as well as when investigating any real physical process, a necessity arises to select the most influential factors from the variety which exist in order to describe the process to a reasonable first approximation. In the extratropical atmosphere, for large-scale processes, such a first approximation does exist and this is the adiabatic approximation, as the following naive arguments suggest. It is clearly visible on weather maps and the Earth's cloudy cover images from satellites, that the large-scale atmospheric phenomena and features may transform very quickly, with the characteristic time-scale of 1–2 days, and it is well understood that such rapid metamorphoses cannot be explained directly by external forces, notably radiative heating and friction. Note that the 'fastest' factor of this sort, namely the friction between the atmosphere and the underlying Earth's surface, has a noticeably greater characteristic time-scale of about 6–7 days.

The adiabatic approximation has a property of primary theoretical and practical significance, namely it allows a set of *conservation laws*: energy, entropy, angular momentum, potential vorticity, and so on. These quantities are named the constants of adiabatic motion, or simply *adiabatic invariants*. This book aims to give a comprehensive and self-consistent account of up-to-date efforts to apply adiabatic invariants in the field of large-scale atmospheric dynamics constructively.

Conservation laws of energy and, particularly, of *atmospheric angular momentum* bear the most classical character and have the longest history of use in dynamic meteorology. The thermodynamic principle of specific entropy material conservation has been successfully used in meteorology since the beginning of the 20th century and then put into the basis of the method of *isentropic analysis* (see Chapter 4). Having no analogs in classical mechanics, the principle of substantial conservation of *potential vorticity* was discovered by C.-G. Rossby (1898–1957) in the late 1930s for a particular case of layered fluid currents, the most general and rigorous formulation being given by H. Ertel (1904–1971) in 1942. Nowadays, the potential vorticity conservation law is a cornerstone of atmospheric dynamics. Traditional quasi-geostrophic governing equations (see also Phillips, 1963), which are substantially due to J. G. Charney and A. M. Obukhov (1918–1989) and govern nearly geostrophic large-scale atmospheric flows, have the same complete set of adiabatic invariants, including *quasi-geostrophic potential vorticity* (sometimes referred to as *pseudo-potential vorticity*) conservation law, as the primitive, non-reduced, governing equations possess. Definite 'intermediate' approximations to primitive governing equations exist, which are to a certain extent more accurate than quasi-geostrophic equations (see, e.g., Salmon, 1998) but they are less universal (not so robust) and more difficult to solve. This is the reason why only traditional quasi-geostrophic equations are systematically exploited in Chapters 2, 3 and 5.

According to *Noether's theorem*, basic space–time symmetries are directly related to the conservation laws of the governing equations. The unique standing fluid *particle-relabeling symmetry* property corresponds to Ertel's (1942) potential vorticity theorem, as is nicely explained in Salmon (1998). Despite this issue being extremely interesting and fundamental, the closely related Lagrangian viewpoint on fluid dynamics will never be used in this book and more 'pragmatic' Eulerian approach is chosen.

One terminology remark seems to be necessary. In classical Hamiltonian mechanics, *adiabatic invariants* are said be the quantities which remain constant when the parameters entering the Hamiltonian function very slowly, or *adiabatically*, change in time, as is explained in ter Haar (1964) and Landau and Lifshitz (1973). Classical statistical mechanics postulates that the total entropy of the system is well-specified by the distribution of the system's components between allowed energy states. During the slow, or adiabatic, time changes in parameters, any transition from one such state to another is restricted, and the total entropy must be kept constant. The latter corresponds well to the adiabatic approximation in thermodynamics, with conservation of thermodynamical entropy.

The analogy between adiabatic invariants in atmospheric large-scale dynamics and those in the frame of classical Hamiltonian mechanics is appealing. Consider, for a while, slowly evolving in time, diabatically and frictionally forced large-scale atmospheric motions which are organized in such a way that the external factors in question nearly annihilate the action of each other, namely the net force is much less than each component taken separately. For such very slow, quasi-equilibrium, atmospheric climate processes, adiabatic invariants could continue to exist and might serve as informative constraints.

The majority of planets in the solar system, and some of their largest satellites, have gaseous envelopes, or atmospheres. When studying the atmospheric general circulation on a variety of nebular objects, especially on the Earth-like Venus and Mars, one might expect to gain a better understanding of atmospheric processes on our home planet, particularly because this offers the unique possibility to verify the basic fluid-dynamic and thermodynamic principles in a broader range of control parameter changes than is ever achievable in the terrestrial atmosphere, at least, in its current climate state. An illuminating example of opening prospects of this sort was an attempt to extrapolate the physical conditions accompanying 'global dust storms' in the Martian atmosphere onto a prediction of global cooling in the Earth's atmosphere, namely the 'nuclear winter', as the result of a major anthropogenic impact on it (see Boubnov and Golitsyn, 1995).

Similar words may be attributed to oceanography and the dynamics of the Sun's atmosphere. Despite striking differences in physical properties between marine water, solar plasma and atmospheric air, it is surprising that there is a lot in common between large-scale dynamics in the Ocean, terrestrial and solar atmospheres. It is interesting that such eminent experts in dynamic meteorology, as V. Bjerknes (1862–1951), C.-G. Rossby and V. Starr (1909–1976), were deeply interested in the Sun's atmosphere dynamics and made an important contribution to the subject. A lot can be said for how detailed knowledge of a neighbouring geophysical science and, moreover, personal experience of working in it could enrich the intuition of a meteorologist and an oceanographer and help them in their own specific research field. Numerous examples from the history of our science, the life and scientific career of C.-G. Rossby maybe one of the brightest (see Bolin, 1999 and references therein), well confirm the last assertion.

A number of reasons and motivations, including some of those mentioned above, facilitated the origination of a joint science, *Geophysical Fluid Dynamics* (GFD), and dynamic meteorology, or more specifically large-scale atmospheric dynamics, enters GFD now as an essential constitutive part. Certainly, the forthcoming chapters are addressed to meteorological fluid dynamicists, in the first place. Nonetheless, the author would be glad if his colleagues from a broader GFD community will find them interesting, too.

CHAPTER 1

Equations of Motion and Conservation Laws

1.1 Equations of compressible fluid dynamics

Atmospheric motion is the flow of a compressible medium, i.e., the air, the physical properties of which are close to those of a perfect gas. Following A. A. Friedmann (1934), the dynamic equations for a compressible fluid (gas) can be divided into two groups. The first group, which we shall call the dynamic group, includes both Euler equations to account for forces which act on fluid parcels and the continuity equation. The second, thermodynamic, group consists of the energy equation, the equation of state and those additional relations between dependent variables which appear when writing the heating terms.

In a chosen frame of reference the dynamic group of equations can be written as follows

$$\frac{D}{Dt}\mathbf{v} = -\frac{1}{\rho}\nabla p + \mathbf{F} + \frac{1}{\rho}\mathbf{F}', \tag{1}$$

$$\frac{\partial}{\partial t}\rho + \nabla \cdot (\rho \mathbf{v}) = 0. \tag{1'}$$

Here $\mathbf{v}$ is the velocity vector, p is pressure, ρ is the fluid density; $D/Dt = \partial/\partial t + \mathbf{v}\cdot\nabla$ is the symbol of a material time-derivative; $\mathbf{F}$ are the mass forces, having the property that the magnitude of such a force acting on a fluid parcel is proportional to its mass; $\mathbf{F}'$ are the bulk forces with the strength proportional to the volume of a parcel to which they are applied.

The mass forces can be of three types: (i) potential forces $\mathbf{F}_1$ satisfying the equation $\nabla\times\mathbf{F}_1 = 0$, (ii) inertial forces $\mathbf{F}_2$ which appear when a non-Galilean frame of reference is considered, and (iii) some other non-potential forces $\mathbf{F}_3$.

In practical applications of compressible fluid dynamics one customarily deals with the potential gravity force. In the field of atmospheric dynamics the Coriolis force becomes the most important inertial force. The reason is that the centrifugal force, having a potential, can be incorporated into the gravity force as its component. When $\vec{\Omega}$ is the constant vector of the Earth's rotation angular velocity, the Coriolis force is determined by the relation $\mathbf{F}_2 = \mathbf{v} \times 2\vec{\Omega}$. The proof of this statement is given in numerous textbooks on basic mechanics, fluid mechanics and dynamic meteorology, e.g., in Haltiner and Martin, 1957; Goldstein, 1980; Pedlosky, 1987; Salmon, 1998. Non-potential mass forces relatively rarely occur in practice and are not considered hereafter. Nevertheless, many results presented in the monograph are valid for a fluid motion subjected to mass forces of a general type.

The set of bulk forces $\mathbf{F}'$, which act on moving fluid parcels, should include the internal friction force, or that of fluid viscosity. This force is mainly due to velocity shears and is called the Newtonian viscosity force. Using the notation $\mathbf{R}$, we write this force as

$$\mathbf{R} = \eta \nabla^2 \mathbf{v} + \left(\frac{1}{3}\eta + \varsigma \right) \nabla(\nabla \cdot \mathbf{v}),$$

where η is the dynamic shear viscosity coefficient and ς is the bulk viscosity coefficient (Landau and Lifshitz, 1988). Using a well-known formula of vector analysis, it is easy to obtain

$$\mathbf{R} = \left(\frac{4}{3}\eta + \varsigma \right) \nabla(\nabla \cdot \mathbf{v}) - \eta \nabla \times \nabla \times \mathbf{v}.$$

Neglecting the fluid compressibility, which is appropriate in slow vortex motion studies, we obtain a simple approximate formula $\mathbf{R} \approx -\eta \nabla \times \nabla \times \mathbf{v}$.

A characteristic feature of Newtonian viscosity force is that it vanishes for a fluid solid-body-like rotation, having the velocity field $\mathbf{u} = \vec{\omega} \times \mathbf{x}$ where $\vec{\omega}$ is an arbitrary constant vector and $\mathbf{x}$ is the position vector of a fluid parcel. This follows directly from $\nabla \times \mathbf{u} = 2\vec{\omega}$ and $\nabla \times \nabla \times \mathbf{u} \equiv 0$. Moreover, the field $\mathbf{u}$ is non-divergent, $\nabla \cdot \mathbf{u} \equiv 0$. That is why the given expression for $\mathbf{R}$ is equally valid both in inertial (Galilean) and uniformly rotating frames of reference if only in the latter case $\mathbf{v}$ stands for relative velocity. In certain situations, e.g., in the case of shallow water currents, it is also helpful to introduce an external friction force. This force $\mathbf{R}'$ is assumed to have a magnitude proportional to that of velocity but act in opposite direction:

$$\mathbf{R}' = -\eta'$$

Here, η' is the external friction coefficient, sometimes referred to as the Rayleigh friction coefficient. In a more correct way, one has to re-write this formula as

$$\mathbf{R}' = -\eta'\left(\mathbf{v} - \mathbf{u}_0\right),$$

where $\mathbf{u}_0$ is the translation velocity of an underlying rigid surface, measured in the same frame of reference. External friction is essential when studying the stability of large-scale atmospheric motions (see Dolzhanskii *et al.*, 1990, 1992). Related topics will be briefly touched in Chapter 5.

Recent papers on numerical modeling of geophysical fluid flows (e.g., Juckes and McIntyre, 1987; Larichev and Fedotov, 1988) often apply an artificial viscosity of the type

$$\mathbf{R}_N = -\eta_N\left(\underbrace{\nabla\times...\times\nabla}_{2N} \times \mathbf{v}\right),$$

($N = 2, 3, ...$). In the course of the application of such a procedure, fine-scale motions are effectively damped but large-scale spatial modes remain virtually untouched.[1]

The continuity equation (1') can be re-written in the form

$$\frac{D}{Dt}\rho + \rho\,\nabla\cdot\mathbf{v} = 0. \tag{2}$$

A fluid is called fully incompressible if $D\rho/Dt = 0$, i.e., the density of the fluid is a material constant. From Equation (2) it follows that $\nabla\cdot\mathbf{v} = 0$, i.e., $\mathbf{v}$ becomes a solenoidal vector.

Let us turn to equations of the thermodynamical group. We use the first law of thermodynamics which states that energy and heat are equivalent and can convert into each other. When a fluid parcel of unit mass is in local thermodynamic equilibrium and is supplied very slowly by small heat portions δq, this amount of heat is consumed, first, for the increase of internal energy De and, second, for the work of fluid thermal expansion

---

[1] This procedure is also referred to as 'modified dissipativity' or 'hyperviscosity' (see Frisch, 1998). In practical terms, it allows one to overcome partially the deficiencies caused by the neglect of smallest spatial scales, not properly resolved in any numerical atmospheric model; sometimes, it also makes it easier to obtain theoretical results for dissipative atmospheric flows, i.e., attractor dimension estimation, otherwise not easily achievable (Dymnikov and Filatov, 1997).

against the pressure forces, δa: $\delta q = De + \delta a$. Assuming the thermal expansion process to be isotropic, from the Pascal's law it follows that $\delta a = pD\alpha$, where $\alpha = \rho^{-1}$ is the specific volume. If we introduce the specific, per unit mass, entropy s, such that $\delta q = TDs$, where T is the kinetic (absolute) temperature, we can write

$$T\frac{Ds}{Dt} = \frac{De}{Dt} + p\frac{D\alpha}{Dt}.$$

In what follows, it is more convenient to pass from α to pressure p as a new variable and to introduce the specific enthalpy $h = e + p\alpha$. With the help of the latter, we use the second law of thermodynamics in the form[2]

$$T\frac{Ds}{Dt} = \frac{Dh}{Dt} - \alpha\frac{Dp}{Dt}.$$

Further on, we re-write this equation as follows

$$T\frac{Ds}{Dt} = \left(\frac{\partial h}{\partial T}\right)_p \frac{DT}{Dt} + \left[\left(\frac{\partial h}{\partial p}\right)_T - \alpha\right]\frac{Dp}{Dt}.$$

Here, $c_p = (\partial h/\partial T)_p$ is the specific heat at constant pressure and $(\partial h/\partial p)_T - \alpha = T(\partial s/\partial p)_T$. We use the specific Gibbs' potential (the free enthalpy) $g = h - Ts$, such that $Dg = -sDT + \alpha Dp$. Equalizing the mixed second derivatives $\partial^2 g/\partial T\partial p$ and $\partial^2 g/\partial p\partial T$ with each other, we obtain $(\partial s/\partial p)_T = -(\partial \alpha/\partial T)_p$. Based on this, we finally write down

$$\rho T\frac{Ds}{Dt} = \rho c_p\frac{DT}{Dt} - \beta T\frac{Dp}{Dt}, \tag{3}$$

where $\beta = -\rho^{-1}(\partial \rho/\partial T)_p$ is the coefficient of fluid thermal expansion.

Fluid diabatic heating at a rate Q per unit volume results in the change of entropy s according to the equation

---

[2] Just after the entropy s has been introduced, we pass immediately from the first to the second law of thermodynamics, the latter taken in its simplest (trivial) form valid for thermodynamically reversible processes only.

$$\rho T \frac{Ds}{Dt} = Q \tag{4}$$

with the right-hand-side term, which is helpful to represent as

$$Q = Q' + \nabla \cdot (\lambda \nabla T).$$

Here, λ is the fluid thermoconductivity coefficient and the term Q' includes both the external diabatic heating (in the atmosphere, this is primarily due to the absorption of solar radiation) and the warming due to the viscous dissipation of the kinetic energy of fluid motion. By combining Equations (3, 4), we shall have

$$\rho c_p \frac{DT}{Dt} - \beta T \frac{Dp}{Dt} = Q. \tag{5}$$

Let us consider the case of the atmospheric air. To close the system of governing equations it is necessary to invoke an equation of state. This is a functional dependence between thermodynamical variables which does not include time explicitly and, thus, characterizes a fluid during the entire period of its motion. For a baroclinic fluid with two independent variables of state the equation we are looking for can be written in a general form either as $f(p,\rho,T) = 0$ or $f'(p,\rho,s) = 0$, where f and f' are known functions. Many results to be obtained later are valid for a general form of the equation of state. Nevertheless, keeping in mind the desirable simplicity of the formulae, we assume that the equation of state is given by the Clapeyron–Mendeleyev equation

$$p = \rho RT. \tag{6}$$

Here, R is the gas constant of the air, which is assumed to be dry, with a constant mixing ratio of the gaseous components. Now, Equation (5) takes the form

$$c_p \frac{D \ln T}{Dt} - R \frac{D \ln p}{Dt} = \frac{Q}{\rho T}. \tag{7}$$

Equations (1, 2, 6, 7) form a closed system of governing equations for atmospheric dynamics provided the heating rate Q is specified.

In liquids, because of their poor compressibility, the second left-hand-side term in Equation (5) is very small compared to the first term. So, with good accuracy

$$\rho c_p \frac{DT}{Dt} \approx Q.$$

For most liquids (one exception is ordinary water at temperatures close to 4°C), $\rho - \rho_0 \approx -\beta\rho_0(T - T_0)$, where ρ_0 is the density of a liquid at standard temperature T_0 and $\beta > 0$. For this reason, we write, instead of Equation (7),

$$\frac{D\rho}{Dt} \approx -\left(\frac{\beta\rho_0}{\rho c_p}\right)Q.$$

In the absence of heating $D\rho/Dt \cong 0$. The condition $\nabla\cdot\mathbf{v} = 0$ of velocity solenoidality replaces the continuity equation (2) for liquids.[3]

In the meteorology, the potential temperature $\theta = T(p_{00}/p)^k$ concept is preferred instead of entropy s. Here, $k = R/c_p$ and $p_{00} = 10^3$ hPa. The potential temperature is in one-to-one correspondence to the specific entropy $s = c_p\ln\theta + s_0$, where the exact value of an additive constant s_0 is not essential because s_0 drops out after taking derivatives. In terms of θ, Equation (7) could be re-written in a more concise form as

$$c_p \frac{D\ln\theta}{Dt} = \frac{Q}{\rho T}.$$

By definition, θ is equal to that temperature T, which an air parcel would attain if it were replaced adiabatically, i.e., without the heat exchange with the ambient air, from its initial baric level p to the reference pressure level of 10^3 hPa. For a perfect gas, the dynamic equations governing the fluid motion could be written in a form more beneficial for applications, with the help of θ and the Exner function $\Pi = c_p(p/p_{00})^k$ as a pair of dependent thermodynamic variables (Eliassen and Kleinschmidt, 1957). Equations (1, 2, 7) become

$$D\mathbf{v}/Dt = -\theta\nabla\Pi - \nabla\Phi + \mathbf{v} \times 2\vec{\Omega} + \mathbf{F},$$

$$\frac{D\Pi}{Dt} + (\kappa-1)\Pi\,\nabla\cdot\mathbf{v} = (\kappa-1)\frac{\hat{Q}}{\theta},$$

$$D\theta/Dt = \hat{Q}/\Pi.$$

---

[3] This is a nearly incompressible fluid approximation (see, e.g., Müller, 1995).

Here, $\hat{Q} = Q/\rho$ is the specific, per unit mass, diabatic heating rate, Φ is the gravity potential, $\mathbf{F}$ stands for non-potential forces including viscosity, $\kappa = c_p/c_v$ is the ratio of specific heats at constant pressure and constant volume, respectively.

With the account for the second law of thermodynamics, we write Equation (1) in the following form

$$\frac{D\mathbf{v}}{Dt} + 2\vec{\Omega} \times \mathbf{v} = -\nabla h + T\,\nabla s - \nabla\Phi + \mathbf{F}.$$

It is convenient to re-write this equation in the so-called Gromeka–Lamb form (see Landau and Lifshitz, 1988)

$$\frac{\partial \mathbf{v}}{\partial t} + \nabla\left(\frac{\mathbf{v}^2}{2}\right) + \vec{\omega}_a \times \mathbf{v} = -\nabla h + T\,\nabla s - \nabla\Phi + \mathbf{F}, \qquad (8)$$

where $\vec{\omega}_a = \nabla \times \mathbf{v} + 2\Omega$ is the absolute vorticity, i.e., the vorticity measured in an absolute, or inertial, frame of reference. If we introduce the Bernoulli function $B = (\mathbf{v}^2/2) + h + \Phi$, Equation (8) could be written in a shorter form

$$\frac{\partial \mathbf{v}}{\partial t} + \vec{\omega}_a \times \mathbf{v} = -\nabla B + T\,\nabla s + \mathbf{F}.$$

We take the curl of this equation

$$\frac{\partial}{\partial t} \nabla \times \mathbf{v} + \nabla \times \left(\vec{\omega}_a \times \mathbf{v}\right) = \nabla T \times \nabla s + \nabla \times \mathbf{F},$$

and, using the well-known vector analysis formulae, arrive at the general equation of three-dimensional vorticity vector transformation, which is called the Friedmann equation in the Russian language literature[4]

$$\frac{D\vec{\omega}_a}{Dt} - \left(\vec{\omega}_a \cdot \nabla\right)\mathbf{v} + \vec{\omega}_a(\nabla \cdot \mathbf{v}) = \nabla T \times \nabla s + \nabla \times \mathbf{F}. \qquad (9)$$

---

[4] This and further results have been established by a Russian mathematician and physicist A. A. Friedmann in his PhD paper in 1922 and published as a monograph in 1934. Friedmann is a world-known person for his seminal contribution to cosmology. He was the first to derive a non-stationary solution for an expanding Universe in the frame of Einstein's general relativity theory.

A linear (by velocity **v**) operator acting on the $\bar{\omega}_a$-field, standing in this equation on left-hand side was named by Friedmann (1934) the Helmholtzian (helm) in honour of H. von Helmholtz. For an arbitrary vector field **A**(**x**,*t*) one has, by definition,

$$\text{helm}\mathbf{A} = \frac{D}{Dt}\mathbf{A} - (\mathbf{A} \cdot \nabla)\mathbf{v} + \mathbf{A}(\nabla \cdot \mathbf{v}).$$

In Friedmann (1934) the so-called shortened Helmholtzian

$$\text{Helm}\mathbf{A} = \frac{D}{Dt}\mathbf{A} - (\mathbf{A} \cdot \nabla)\mathbf{v}$$

has been introduced and a remarkable vector identity has been proved

$$\frac{D}{Dt}(\mathbf{A} \cdot \nabla f) = \mathbf{A} \cdot \nabla\left(\frac{Df}{Dt}\right) + \text{Helm}\mathbf{A} \cdot \nabla f,$$

where $f(\mathbf{x}, t)$ stands for an arbitrary scalar function. In order to give a proof, it is sufficient to write down a chain of equalities

$$\frac{D}{Dt}(\mathbf{A} \cdot \nabla f) = \mathbf{A} \cdot \frac{D}{Dt}\nabla f + \frac{D\mathbf{A}}{Dt} \cdot \nabla f$$

$$= \mathbf{A} \cdot \frac{\partial}{\partial t}\nabla f + \mathbf{A} \cdot \left((\mathbf{v} \cdot \nabla)\nabla f\right) + \frac{D\mathbf{A}}{Dt} \cdot \nabla f$$

$$= \mathbf{A} \cdot \nabla\frac{\partial f}{\partial t} + \mathbf{A} \cdot \nabla(\mathbf{v} \cdot \nabla f) - \mathbf{A} \cdot \nabla(\mathbf{v} \cdot \nabla f)$$

$$+ \mathbf{A} \cdot \left((\mathbf{v} \cdot \nabla)\nabla f\right) + \frac{D\mathbf{A}}{Dt} \cdot \nabla f$$

$$= \mathbf{A} \cdot \nabla\left(\frac{Df}{Dt}\right) - ((\mathbf{A} \cdot \nabla)\mathbf{v}) \cdot \nabla f + \frac{D\mathbf{A}}{Dt} \cdot \nabla f$$

$$= \mathbf{A} \cdot \nabla\left(\frac{Df}{Dt}\right) + \text{Helm}\mathbf{A} \cdot \nabla f.$$

It is easier to perform a transition to the fourth equality sign if tensorial notations are used. When the continuity equation (2) is applied, it is easy to arrive at a formula helpful for subsequent calculations (Monin, 1990)

$$\frac{D}{Dt}\frac{\mathbf{A}\cdot\nabla f}{\rho} = \frac{\mathbf{A}\cdot\nabla\left(Df/Dt\right)}{\rho} + \frac{\text{helm}\mathbf{A}\cdot\nabla f}{\rho}. \tag{10}$$

Friedmann (1934) proved a general theorem that helm$\mathbf{A}$ = 0 is the necessary and sufficient condition for the conservation of both vector lines and vector tubes of an arbitrary field $\mathbf{A}(\mathbf{x}, t)$.

When taken in their general form, the governing equations of atmospheric dynamics are extremely complicated for handling. So, one needs to incorporate certain approximations into them. Let us outline three such approximations, as the most physically justified and widely used: (i) adiabatic approximation, (ii) approximation of atmospheric air weak dynamic compressibility (also used in the form of Boussinesq and anelastic approximations (see, e.g., Zeytounian, 1991)), and (iii) quasi-static approximation.

Modern theoretical studies of large-scale atmospheric dynamics widely use a quasi-geostrophic approximation which, however, is essentially more restrictive as compared with the above approximations.

At the very end of this section, it is necessary to say a few words about the possible corrections in the atmospheric governing equations that might appear due to the account for air humidity. An analogy between moist air and sea salt water dynamics is rather often stressed in the literature. Due to a high extent of solubility of different salts in marine water, changes in their concentration along with the temperature fluctuations directly control the dynamic oceanic processes via induced perturbations of water density. This is the essence of the thermohalinity effect. In a moist air the situation is quite different. The partial pressure of water vapor is only 0.1–1% of the net air pressure, and the direct impact of the air moisture on atmospheric dynamics (via corresponding density perturbations, the moist air being a little lighter than the dry air) is not very significant, except in the equatorial area. To be more accurate, this impact is usually smaller than the contribution of temperature fluctuations. So, the motion of a non-saturated moist air can be treated as that of an effectively dry air, without a noticeable loss of accuracy. The situation drastically changes when the moist air reaches its saturation state and the effect of phase transition arises. This leads to the release or, the other way round, absorption of the latent heat of evaporation. This essentially transforms the atmospheric thermodynamic parameters and influences the dynamics through their changes. Here, Equations (1, 2) could be left unchanged with the accuracy sufficient for most applications, but instead of Equation (7) one could write

$$c_p \frac{D \ln T}{Dt} - R \frac{D \ln p}{Dt} = \frac{Q}{\rho T} - \Gamma \frac{Dm}{Dt},$$

where m is the specific humidity of a saturated water vapor and Γ is equal to the ratio of the latent heat of evaporation $L(T)$ and temperature T. In the case of a non-saturated moist air, $Dm/Dt = 0$, and Equation (7) is valid again.

1.2 Adiabatic approximation. Potential vorticity theorem

As it is appropriate in physics, a quantity c is referred to as conservative (e.g., c is energy) if the following equation holds

$$\frac{\partial}{\partial t} c + \nabla \cdot \mathbf{j} = 0, \tag{1}$$

where $\mathbf{j} = \mathbf{v}c + \mathbf{N}_c$ and $\mathbf{N}_c$ is the non-advective part of the total flux of the quantity c. The divergent form of Equation (1) excludes the existence of internal sources and sinks of c. Using Equation (2) from Section 1.1, Equation (1) can be re-written in equivalent terms as

$$\frac{D}{Dt} \left(\frac{c}{\rho} \right) = -\frac{1}{\rho} \nabla \cdot \mathbf{N}_c. \tag{2}$$

When $\mathbf{N}_c \equiv 0$,

$$\frac{D}{Dt} \hat{c} = 0, \quad \hat{c} = \frac{c}{\rho}, \tag{3}$$

and in this particular case $\hat{c}$ is referred to as a material constant. If Equation (1) is integrated over a volume V fixed in space and bounded by an impermeable surface, ∂V, then according to Gauss–Ostrogradsky's theorem it is easy to obtain

$$\frac{d}{dt} \iiint_V c \, d\tau = -\iint_{\partial V} \mathbf{N}_c \cdot \mathbf{n} \, d\sigma, \tag{4}$$

where $\mathbf{n}$ is a unit vector at the surface ∂V directed outward from the volume

V. If the right-hand side of Equation (4) vanishes, we arrive at an integral conservation law

$$\frac{d}{dt}C = 0, \quad C = \iiint_V c d\tau. \tag{5}$$

It is also possible to consider a material volume V as an integration domain. In this case, it is more convenient to start directly from Equation (2). By integrating the latter, we arrive at Equation (4) again, and, consequently, at the conservation law (5), if the right-hand side of Equation (4) vanishes. In the general case, when integration is extended over a volume $V(t)$ which may change due to its boundary surface motion with velocity $\mathbf{u}(\mathbf{x},t)$, then for any function $c(\mathbf{x},t)$ ($\mathbf{x}$ is the position vector) we have the Leibnitz's theorem[5]

$$\frac{d}{dt}\iiint_V c d\tau = \iiint_V \frac{\partial c}{\partial t} d\tau + \iint_{\partial V} c\mathbf{u}\cdot\mathbf{n}d\sigma.$$

Assuming c to be a conservative quantity, we start from Equation (1), apply the Gauss–Ostrogradsky's theorem in intermediate calculations, and finally obtain

$$\frac{d}{dt}\iiint_V c d\tau = -\iiint_V (\nabla\cdot\mathbf{j})d\tau + \iint_{\partial V} c\mathbf{u}\cdot\mathbf{n}d\sigma$$

$$= -\iint_{\partial V}\mathbf{j}\cdot\mathbf{n}d\sigma + \iint_{\partial V} c\mathbf{u}\cdot\mathbf{n}d\sigma = \iint_{\partial V} c(\mathbf{u}-\mathbf{v})\cdot\mathbf{n}d\sigma - \iint_{\partial V}\mathbf{N}_c\cdot\mathbf{n}d\sigma.$$

The first right-hand side term vanishes in two cases mentioned above: (i) an impermeable boundary surface, ∂V, fixed in space, when $\mathbf{v} = \mathbf{u} = 0$, and (ii) a material surface ∂V, with $\mathbf{v} = \mathbf{u}$.

Further on, an adiabatic approximation is considered when $Q = 0$ in Equation (4) of Section 1.1. As one diabatic heating component is due to viscous dissipation, a fluid has to be ideal in order for the adiabatic approximation to hold.

The existence of conservation laws is the fundamental property of adiabatic approximation. Constants of adiabatic motion are named the

---

[5] Here, generally speaking, c can be any rank tensor, not only a scalar, as it is in our case.

adiabatic invariants. They can be either integral (5) or local (3). The latter constants are often referred to as the Lagrangian invariants. In meteorology, one thermodynamical local invariant is well-known for a long time already and is widely used nowadays. It is specific entropy s or, which is equivalent but more convenient, potential temperature θ. Specific humidity m for a motion of non-saturated moist air is also a Lagrangian invariant. Under definite circumstances, the concentration of minor gaseous species is the material constant (3), too. For instance, it could be the ozone concentration, the study of which currently attracts particular attention in connection with the well-known problem of 'ozone depletions' over Antarctic and Arctic regions.

The importance of local invariants of the dynamic origin had been realized in meteorology much later but just they play the central role in it nowadays. A corner-stone is the strict hydrodynamic assertion by Ertel (1942): in an inviscid compressible fluid subjected to potential external forces, when an arbitrary function ψ of pressure p and density ρ obeys Equation (3), the scalar product of the absolute vorticity $\bar{\omega}_a$ on the gradient of ψ divided by the fluid density is the local invariant, i.e., the following conservation law takes place

$$\frac{D}{Dt} \frac{\bar{\omega}_a \cdot \nabla \psi}{\rho} = 0.$$

An important particular case of the application of this theorem is the adiabatic motion of a perfect gas. Here, the Ertel's invariant, following his own pioneering paper, is traditionally expressed in terms of potential temperature θ:

$$\frac{D}{Dt} I = 0, \quad I = \frac{\bar{\omega}_a \cdot \nabla \theta}{\rho}. \tag{6}$$

The invariant I is named Ertel's potential vorticity. In this Chapter, when general compressible fluid dynamics theorems are formulated, we shall set $\psi = s$ because the entropy s has a more general physical content as compared with the potential temperature. In Chapter 4, we shall return to the potential temperature when discussing meteorological implications. In oceanological problems the fluid compressibility can be usually neglected. In this case, the water density ρ replaces the entropy in Ertel's theorem. The proof of Ertel's potential vorticity theorem is given in numerous textbooks on fluid dynamics and geophysical fluid dynamics (e.g., in Kochin *et al.* 1964; Pedlosky, 1987; Landau and Lifshitz, 1988; see also Schröder, 1988).

In the first immediate comment on Ertel's (1942) paper by Moran

(1942) the potential vorticity conservation law was proved illustratively (just in the way which becomes customary nowadays) by using the Lord Kelvin's (W. Tomson) circulation theorem along with the mass conservation principle. Both of them are applied to a material fluid element, 'rolled' within a layer bounded by two closely spaced isentropic surfaces. This illustrative proof has been independently given by Charney (1948). Hereafter, it is reproduced in a slightly modified form proposed by Obukhov (1984). Two neighboring isentropic surfaces with labels s and $s' = s + \delta s$ are taken. Consider a cylinder 'rolled' within a layer between these surfaces, which leans on two reducible closed material contours L and L', lying on the surfaces Σ_s and $\Sigma_{s'}$, respectively. Due to Kelvin's theorem, the velocity circulations $\Gamma = \oint_L \mathbf{v} \cdot d\mathbf{l}$ and $\Gamma' = \oint_{L'} \mathbf{v} \cdot d\mathbf{l'}$ over the curves L and L' are constant following the fluid motion. Here $\mathbf{v}$ is the velocity vector and $d\mathbf{l}$, $d\mathbf{l'}$ are elements of the arc around curves L and L', respectively. Kelvin's theorem is valid because curves L and L' lay in isentropic surfaces, with pressure and density being in one-to-one functional dependence on each other. In this way,

$$d\overline{\Gamma}/dt = 0, \overline{\Gamma} = (\Gamma + \Gamma')/2.$$

Under adiabatic approximation δs is constant, and consequently

$$d[\overline{\Gamma}\delta s]/dt = 0.$$

Along with this the mass of the cylinder is constant. The latter quantity could be taken as

$$\delta M = \rho\sigma h(\delta s),$$

where σ is an average area encircled within the contours L and L', and $h(\delta s)$ is an average distance between Σ_s and $\Sigma_{s'}$ inside the cylinder. That is why

$$\frac{d}{dt}\left[\frac{\overline{\Gamma}\delta s}{\rho\sigma h(\delta s)}\right] = 0.$$

We note that $\delta s = |\nabla s| h(\delta s)$. Using Stokes' theorem and taking the limit case of $\delta s \to 0$, $\sigma \to 0$, one readily arrives at Ertel's potential vorticity I conservation law.

It could be seen that the potential vorticity is a pseudoscalar, i.e., it changes its sign when a mirror transformation of coordinates is performed.

As a result, in the atmospheres over the Northern Hemisphere (NH) and Southern Hemisphere (SH), respectively, the air parcels, which are identical in all other respects, will have l-values of opposite sign. They are positive in the NH and negative in the SH. However, setting of the l-sign is the result of an agreement only and is based on the choice of the positive direction for the vector of the angular velocity of the Earth's rotation.

Under the influence of diabatic heating and non-potential forces, including friction, the potential vorticity transforms according to the equation (Eliassen and Kleinschmidt, 1957; Obukhov, 1962)

$$\frac{D}{Dt}\frac{\vec{\omega}_a \cdot \nabla s}{\rho} = \frac{\vec{\omega}_a \cdot \nabla(Ds/Dt)}{\rho} + \frac{\nabla s \cdot \nabla \times \mathbf{F}}{\rho},\qquad(7)$$

which immediately follows from Equation (10) of Section 1.1 if we put $\mathbf{A} = \vec{\omega}_a, f = s(p, \rho)$ and take into account that Friedmann's equation (9) from Section 1.1. can be identically re-written as

$$\text{helm}\vec{\omega}_a = \rho^{-2}(\nabla\rho \times \nabla p) + \nabla \times \mathbf{F}.$$

In the most general case of $f = \Psi$, where Ψ is an arbitrary function of spatial coordinates and time, we arrive at the formula

$$\frac{D}{Dt}\frac{\vec{\omega}_a \cdot \nabla\Psi}{\rho} = \frac{1}{\rho}\nabla\Psi \cdot \left(\nabla p \times \nabla\left(\frac{1}{\rho}\right)\right) + \frac{\vec{\omega}_a \cdot \nabla(D\Psi/Dt)}{\rho} + \frac{\nabla\Psi \cdot \nabla \times \mathbf{F}}{\rho}.$$

$$(8)$$

In the absence of the last right-hand side term in Equation (8), i.e., for an inviscid fluid subjected to potential external forces only, this formula was discovered by Ertel (1942) and laid into the basis of his proof for the potential vorticity theorem.

The potential vorticity concept enables one to distinguish strictly between two general classes of compressible fluid vortical motion: (i) flows with vanishing potential vorticity, when the vorticity vector is tangent everywhere to isentropic surfaces and, as the result, vorticity filaments lie on these surfaces, and (ii) flows with non-zero potential vorticity, when the vorticity vector penetrates the isentropic surfaces.

Motions of the first class could be introduced correctly only when the right-hand side of Equation (7) vanishes simultaneously. An important example is a two-dimensional, in the vertical plane, flow of a compressible fluid with neglect of the background fluid rotation or, more specifically, when the angular velocity vector of background fluid rotation is horizontal. Air motion in the equatorial atmosphere belongs to this case. Air flows of the first class are frequently met in mesometeorology, and their analysis is important, e.g., for extra-short-term weather forecasts, including the prediction of severe weather events, like whirlwinds, squall-lines, tornadoes, rotating storms, etc.

Large-scale synoptic processes belong to the second class flows. Here, the Earth's background rotation plays a crucial role. Potential vorticity sources and sinks are also important.

1.3   Vorticity charge (potential vorticity substance)

The potential vorticity equation can be derived based on the general theorem for vorticity charge conservation in a compressible fluid (Obukhov, 1962). To do it, we start from Euler's equations (1) of Section 1.1 written in a symbolic form

$$\partial \mathbf{v}/\partial t = \mathbf{f}. \tag{1}$$

Here, $\mathbf{f} = \mathbf{v} \times \vec{\omega}_a - \nabla B + T\,\nabla s + \mathbf{F}$ and all notations are explained in Section 1.1. We take the curl of Equation (1)

$$\frac{\partial}{\partial t} \nabla \times \mathbf{v} = \nabla \times \mathbf{f} \tag{2}$$

and multiply scalarly Equation (2) by an arbitrary vector $\mathbf{A}$ obeying the equation $\partial \mathbf{A}/\partial t = \mathbf{G}$. According to a well-known vector identity, we have

$$\frac{\partial}{\partial t}(\mathbf{A} \cdot \nabla \times \mathbf{v}) = \nabla \cdot (\mathbf{f} \times \mathbf{A}) + \mathbf{f} \cdot \nabla \times \mathbf{A} + \mathbf{G} \cdot \nabla \times \mathbf{v}. \tag{3}$$

Because the fluid background rotation vector $\vec{\Omega}$ is constant, Equation (2) can be identically re-written as

$$\frac{\partial}{\partial t}\left(\nabla \times \mathbf{v} + 2n\vec{\Omega}\right) = \nabla \times \mathbf{f},$$

where n is an arbitrary integer. With account for this, Equation (3) attains a more general form

$$\frac{\partial}{\partial t}\left\{\mathbf{A}\cdot\left(\nabla\times\mathbf{v}+2n\vec{\Omega}\right)\right\}=\nabla\cdot(\mathbf{f}\times\mathbf{A})+\mathbf{G}\cdot\left(\nabla\times\mathbf{v}+2n\vec{\Omega}\right)+\mathbf{f}\cdot\nabla\times\mathbf{A}. \quad (4)$$

Let us consider a special case of the application of the resulting formula. In Equation (4), we put $\mathbf{A}=\nabla\lambda$, $\mathbf{G}=\nabla H$, where λ is an arbitrary scalar field and $\partial\lambda/\partial t=H$. As the result, we have

$$\frac{\partial}{\partial t}\left\{\nabla\lambda\cdot\left(\nabla\times\mathbf{v}+2n\vec{\Omega}\right)\right\}=\nabla\cdot(\mathbf{f}\times\nabla\lambda)+\nabla H\cdot\left(\nabla\times\mathbf{v}+2n\vec{\Omega}\right).$$

Further on, this equation is re-written in the form of the conservative law (1) from Section 1.2

$$\frac{\partial}{\partial t}\left\{\nabla\lambda\cdot\left(\nabla\times\mathbf{v}+2n\vec{\Omega}\right)\right\}+\nabla\cdot\left\{\nabla\lambda\times\mathbf{f}+\nabla H\times\mathbf{v}-2n\vec{\Omega}H\right\}=0. \quad (5)$$

The most important choice for practical applications is that of $n=1$, $\lambda=s(p,\rho)$. Substituting the concrete form of $\mathbf{f}$ into Equation (5), performing some identical transformations and omitting certain non-divergent terms in braces in the left-hand side of Equation (5), one arrives at the equation of vorticity charge conservation (Obukhov, 1962):

$$\frac{\partial}{\partial t}\left(\vec{\omega}_a\cdot\nabla s\right)+\nabla\cdot\mathbf{j}=0. \quad (6)$$

The vector of the total vorticity charge flux $\mathbf{j}$ can be taken in several equivalent forms (Obukhov, 1962; Haynes and McIntyre, 1987; Kurgansky and Tatarskaya, 1987)

$$\mathbf{j}=\mathbf{v}\left(\vec{\omega}_a\cdot\nabla s\right)-\vec{\omega}_a J-s(\nabla\times\mathbf{F})\div\mathbf{v}\left(\vec{\omega}_a\cdot\nabla s\right)-\vec{\omega}_a J-\mathbf{F}\times\nabla s$$

$$\div\mathbf{v}\left(\vec{\omega}_a\cdot\nabla s\right)-2\vec{\Omega}J-\mathbf{v}\times\nabla J-\mathbf{F}\times\nabla s, \quad (7)$$

where ($\div$) denotes the equality with the accuracy of a non-divergent vector, and the notation $J=Ds/Dt=H+\mathbf{v}\cdot\nabla s$ is used.

When a flow is adiabatic and subjected to potential forces only, Equation (6) takes a form of the continuity equation (1) from Section 1.1 with the accuracy of ρ to be replaced by $\vec{\omega}_a\cdot\nabla s$:

$$\frac{\partial}{\partial t}\left(\vec{\omega}_a \cdot \nabla s\right) + \nabla \cdot \left(\mathbf{v}\left(\vec{\omega}_a \cdot \nabla s\right)\right) = 0.$$

Re-writing this equation as

$$\frac{\partial}{\partial t}\left(\rho\frac{\vec{\omega}_a \cdot \nabla s}{\rho}\right) + \nabla \cdot \left(\rho\mathbf{v}\,\frac{\vec{\omega}_a \cdot \nabla s}{\rho}\right) = 0$$

and using the continuity equation, we immediately arrive at the Ertel's potential vorticity conservation theorem $dI/dt = 0$, $I = \left(\vec{\omega}_a \cdot \nabla s\right)/\rho$, the latter quantity being interpreted as the specific vorticity charge density per unit mass. Contrary to the term 'potential vorticity substance' (PVS), adopted by Haynes and McIntyre (1990), we prefer 'vorticity charge' as suggested by Obukhov (1962). Among other things, the latter term stresses an analogy between the fluid dynamic theorem on $\vec{\omega}_a \cdot \nabla s$ conservation and the mathematical expression for electric charge conservation in electrodynamics. As it is appropriate in physics, potential vorticity I is referred to as an intensive quantity, similar in this respect to the kinetic temperature T, and vorticity charge ρI as an extensive, or additive, quantity. The total vorticity charge, contained in a fluid volume V is expressed by the integral

$$Z = \iiint\limits_{V} \vec{\omega}_a \cdot \nabla s\, d\tau.$$

According to Gauss–Ostrogradsky's theorem, as vector $\vec{\omega}_a$ is solenoidal,

$$Z = \iint\limits_{\partial V} \left(\vec{\omega}_a \cdot \mathbf{n}\right) s\, d\sigma,$$

where $\mathbf{n}$ is a unit vector orthogonal to ∂V and directed outward from the volume V. When the domain V can be contracted either into a point or into a closed curve (in the latter case V is a torus-like domain) and is bounded by a closed isentropic surface $s = s_0$,

$$Z = \oiint\limits_{\partial V}\left(\vec{\omega}_a \cdot \mathbf{n}\right)s_0\, d\sigma = s_0\oiint\limits_{\partial V}\left(\vec{\omega}_a \cdot \mathbf{n}\right)d\sigma = 0,$$

and there are exactly equal amounts of positive and negative vorticity charge inside V, which annihilate each other (Haynes and McIntyre, 1987;

McIntyre and Norton, 1990). If we have such a volume V bounded by a closed isentropic surface Σ, and, what is more, there are volumes V_i ($i = 1$, ..., n) which are cut-off from its interior and have non-isentropic closed boundaries σ_i, the net vorticity charge can attain non-zero values. An example is the entire Earth's atmosphere as a spherical shell bounded from inside by the non-isentropic Earth's surface.

However, we could formally apply the vorticity charge conservation theorem to the entire physical volume V, bounded by a closed isentropic envelope Σ, if we assume that inner volumes V_i are filled with a certain weightless substance. It is supposed that a substance inside each volume V_i possesses a certain amount of vorticity charge Z_i, the latter value being quite arbitrary. The only requirement is that the sum of the vorticity charge Z in a real fluid and of virtual vorticity charges Z_i is equal to zero: $Z + \sum_{i=1}^{n} Z_i = 0$.

These speculations become more definite in the case of $n = 1$. The example of the Earth's atmosphere as a spherical gaseous shell is such a case. Our treatment is essentially topological, i.e., it disregards the volume ratio $V^{-1} \sum_{i=1}^{n} V_i$. Tending this ratio to zero, we arrive at the case of volume V with cut-off internal points where the singular vorticity charges are placed. Now, the volume V, being bounded by a closed isentropic surface, can accumulate non-zero vorticity charges. In the nature, it can happen when aerosols and water droplets are suspended in the air.

1.4　Helicity

The generality of Equation (4) of the preceding section enables one to formulate another fundamental conservation principle in fluid dynamics, namely that of helicity, or kinematic helicity, to be distinguished from magnetic helicity in magnetohydrodynamics, determined as a scalar product of the magnetic field on the vector potential (see, e.g., Lesieur, 1997).

In fluid dynamics, the helicity concept appeared for the first time in connection with the problem of the construction of isoscalar surfaces $\Psi(\mathbf{x}) =$ const, which should be orthogonal to fluid streamtubes of finite cross-section (see Loitsyansky, 1973). By the definition of a streamtube, the condition $\mathbf{v} = \lambda(\mathbf{x})\nabla\Psi$ has to hold for an arbitrary function λ. Taking the curl of this equality, we obtain $\nabla\times\mathbf{v} = \nabla\lambda\times\nabla\Psi$ and, by necessity, $\mathbf{v}\cdot\nabla\times\mathbf{v} = 0$. When the quantity $\mathbf{v}\cdot\nabla\times\mathbf{v} = 0$, which is called the helicity, is non-zero, such a construction is impossible. Meanwhile, let us note that an arbitrary vector field $\mathbf{A}(\mathbf{x},t)$ with $\mathbf{A}\cdot\nabla\times\mathbf{A} \neq 0$ is referred to as a screwed (helical) field, and $\mathbf{A}\cdot\nabla\times\mathbf{A}$ is known to be the helicity of the vector field $\mathbf{A}$.

We apply the helicity concept to the velocity vector field in non-stationary fluid flows. Setting $n = 2$, $\mathbf{A} = \mathbf{v}$, $\mathbf{G} = \mathbf{f}$ in Equation (4) of Section 1.3, we obtain

$$\frac{\partial}{\partial t}\left\{\mathbf{v} \cdot \left(\nabla \times \mathbf{v} + 4\vec{\Omega}\right)\right\} = \nabla \cdot \{\mathbf{f} \times \mathbf{v}\} + 2\vec{\omega}_a \cdot \mathbf{f}.$$

In the left-hand side of this equation, under the partial time-derivative sign, one finds a quantity which generalizes the helicity concept onto the case of background fluid rotation. We write this equation using the explicit form of vector $\mathbf{f}$:

$$\frac{\partial}{\partial t}\left\{\mathbf{v} \cdot \left(\nabla \times \mathbf{v} + 4\vec{\Omega}\right)\right\} = \nabla \cdot \left\{\left(\mathbf{v} \times \vec{\omega}_a\right) \times \mathbf{v} - \nabla B \times \mathbf{v}\right.$$

$$+ T\,\nabla s \times \mathbf{v} + \mathbf{F} \times \mathbf{v}\} + 2\vec{\omega}_a \cdot \left(\mathbf{v} \times \vec{\omega}_a\right) + 2T\left(\vec{\omega}_a \cdot \nabla s\right)$$

$$+ 2\vec{\omega}_a \cdot \mathbf{F} - 2\vec{\omega}_a \cdot \nabla B\,.$$

Using the identities

$$\vec{\omega}_a \cdot \left(\mathbf{v} \times \vec{\omega}_a\right) \equiv 0, \quad 2\vec{\omega}_a \cdot \nabla B = \nabla \cdot \left(2\vec{\omega}_a B\right),$$

$$\nabla \cdot \left(\nabla B \times \mathbf{v}\right) = -\nabla \cdot \left(B\,\nabla \times \mathbf{v}\right)$$

and the definition of Ertel's potential vorticity I

$$2T\left(\vec{\omega}_a \cdot \nabla s\right) = 2T\frac{\vec{\omega}_a \cdot \nabla s}{\rho}\rho = 2\,T\,I\rho$$

we arrive at the equation of general helicity balance

$$\frac{\partial}{\partial t}\left\{\mathbf{v} \cdot \left(\nabla \times \mathbf{v} + 4\vec{\Omega}\right)\right\} = -\nabla \cdot \left\{\left(\vec{\omega}_a \times \mathbf{v}\right) \times \mathbf{v}\right.$$

$$+ \left(\nabla \times \mathbf{v} + 4\vec{\Omega}\right)B - T\,\nabla s \times \mathbf{v} - \mathbf{F} \times \mathbf{v}\} + 2T\,I\rho + 2\vec{\omega}_a \cdot \mathbf{F}. \qquad (1)$$

A different procedure for the derivation of this equation was used in Kurgansky (1989) which generalized a more specific relation established by Hide (1989).

Now, we consider a special case of isentropic flow of a compressible fluid ($s = \text{const}$). Here, Equation (1) takes the form

$$\frac{\partial}{\partial t}\left\{\mathbf{v}\cdot\left(\nabla\times\mathbf{v}+4\vec{\Omega}\right)\right\} = -\nabla\cdot\left\{\left(\vec{\omega}_a\times\mathbf{v}\right)\times\mathbf{v}\right.$$

$$+\left(\nabla\times\mathbf{v}+4\vec{\Omega}\right)B - \mathbf{F}\times\mathbf{v}\Big\} + 2\vec{\omega}_a\cdot\mathbf{F}.$$

Using the identity

$$\left(\vec{\omega}_a\times\mathbf{v}\right)\times\mathbf{v} = \mathbf{v}\left(\vec{\omega}_a\cdot\mathbf{v}\right) - \vec{\omega}_a\cdot\mathbf{v}^2,$$

we re-write this equation as

$$\frac{\partial}{\partial t}\left\{\mathbf{v}\cdot\left(\nabla\times\mathbf{v}+4\vec{\Omega}\right)\right\} = -\nabla\cdot\left\{\mathbf{v}\left(\vec{\omega}_a\cdot\mathbf{v}\right) - \left(\nabla\times\mathbf{v}\right)\left(\frac{\mathbf{v}^2}{2}-h-\Phi\right)\right.$$

$$+ 4\vec{\Omega}(h+\Phi) - \mathbf{F}\times\mathbf{v}\Big\} + 2\vec{\omega}_a\cdot\mathbf{F}.$$

It is possible to simplify the resulting equation in the case when there is no background fluid rotation $\left(\vec{\Omega} = 0\right)$ and only potential forces are present $(\mathbf{F} = 0)$

$$\frac{\partial}{\partial t}(\mathbf{v}\cdot\nabla\times\mathbf{v}) = -\nabla\cdot\left\{\mathbf{v}(\mathbf{v}\cdot\nabla\times\mathbf{v}) - L\nabla\times\mathbf{v}\right\},$$

$$L = \frac{1}{2}\mathbf{v}^2 - h - \Phi.$$

We re-write this equation as

$$\frac{D}{Dt}(\mathbf{v}\cdot\nabla\times\mathbf{v}) + (\mathbf{v}\cdot\nabla\times\mathbf{v})(\nabla\cdot\mathbf{v}) = \nabla\cdot(L\nabla\times\mathbf{v})$$

and eliminate the velocity divergence $\nabla\cdot\mathbf{v}$ using the continuity equation. As the result, we have

$$\frac{D}{Dt}\frac{\mathbf{v}\cdot\nabla\times\mathbf{v}}{\rho}=\frac{1}{\rho}\nabla\cdot(L\nabla\times\mathbf{v}). \tag{2}$$

Integration of Equation (2) over a material volume V gives

$$\frac{d}{dt}\iiint_V(\mathbf{v}\cdot\nabla\times\mathbf{v})d\tau=\iiint_V\nabla\cdot(L\nabla\times\mathbf{v})d\tau=\iint_{\partial V}(\mathbf{n}\cdot\nabla\times\mathbf{v})Ld\sigma,$$

where $\mathbf{n}$ is a unit vector orthogonal to ∂V and directed outward the volume V. When $\mathbf{n}\cdot\nabla\times\mathbf{v}\equiv 0$, i.e., no vorticity filament penetrates the boundary surface, the invariant of helicity exists (Moreau, 1961; Moffatt, 1969):

$$H=\iiint_V(\mathbf{v}\cdot\nabla\times\mathbf{v})\,d\tau=\text{invariant}.$$

The invariant H is a measure of structural complexity of the velocity field. For example, when there are singular vortex filaments inside volume V, H characterizes the degree of their knottedness or linkage (Moffatt, 1969). In the simplest case of two linked concentrated vortex filaments with intensities Γ_1 and Γ_2, one has $H=\pm 2\Gamma_1\Gamma_2$. It is possible to prove this statement using quite simple arguments. The invariant H is of topological origin and its numerical value does not change at any continuous deformation of vortex filaments. Let us stretch one of them to such an extent that it can be considered locally as a straight vortex filament closed at infinity. The second vortex filament is transformed to the circular vortex ring of radius R_2 with the center posed in any point of the straight vortex filament. The vortex ring should be in a plane orthogonal to the straight vortex filament. According to the Biot-Savart's law, the tangent velocity induced by the straight vortex filament at distance R_2 is equal to $v_1=\Gamma_1/2\pi R_2$ and is directed according to the right-hand screw rule. As a result, the corresponding contribution to H is equal to $\pm 2\pi R_2 v_1\Gamma_2=\pm\Gamma_1\Gamma_2$. Here, the plus sign corresponds to the case when the vortex filaments belong to the right-hand screw system, i.e., to the case when the circular vortex ring moves along the direction of the vorticity vector for the straight vortex filament. The minus sign agrees with the alternative case. Actually, the vortex filaments enjoy equal rights. Replacing them, i.e., stretching the second vortex filament and making the first one circular, of radius R_1, we shall have the contribution $\pm 2\pi R_1(\Gamma_2/2\pi R_1)\Gamma_1=\pm\Gamma_2\Gamma_1$ to helicity H. Summing up, one has $H=\pm\Gamma_1\Gamma_2\pm\Gamma_2\Gamma_1=\pm 2\Gamma_1\Gamma_2$ which completes the proof.

In a non-isentropic fluid flow the helicity H loses its invariant properties. Instead, it evolves in time according to Equation (1). If one restricts himself to motions with vanishing Ertel's potential vorticity $I \equiv 0$, when all non-potential forces, including viscous friction, are also absent (see Section 1.2), the helicity becomes the constant of motion again.

The existence of the second, supplementary to energy, non-trivial constant of motion essentially transforms the properties of the barotropic fluid flows and grants a relative persistence to helical flow structures. Examples are tornadoes, whirlwinds, horizontally oriented vortex-like structures in the planetary boundary layer ('cloud streets'), etc. (Etling, 1985; Lilly, 1985).

Using the Lagrangian action W defined by the formula $L = DW/Dt$, Equation (2) can be re-written as

$$\frac{D}{Dt} \frac{\mathbf{v} \cdot \nabla \times \mathbf{v}}{\rho} = \frac{1}{\rho} \nabla(DW/Dt) \cdot \nabla \times \mathbf{v}.$$

Recalling Ertel's general formula (8) from Section 1.2, we put $\Psi = W$ in it and take into account that $\nabla p \times \nabla \rho = 0$ for a barotropic fluid. Immediately, we have

$$\frac{1}{\rho} \nabla\left(\frac{DW}{Dt}\right) \cdot \nabla \times \mathbf{v} = \frac{D}{Dt} \frac{\nabla W \cdot \nabla \times \mathbf{v}}{\rho}$$

and, thus, arrive to the material conservation law

$$\frac{D}{Dt} \frac{(\mathbf{v} - \nabla W) \cdot \nabla \times \mathbf{v}}{\rho} = 0, \tag{3}$$

discovered by Ertel and Rossby (1949). This material constant has served as a predecessor for a set of Hollmann's (1964) material constants of the three-dimensional adiabatic flow of a baroclinic fluid. Hollmann's invariants have not found broad practical applications yet, though they are regularly discussed in the literature (see, e.g., Diky, 1972; Egger, 1989). Here, we note that equation (3), along with the above additional constrains on the topological structure of the vorticity field, immediately results in the conservation of helicity H.

Meso- and small-scale atmospheric vortices, e.g., tornadoes, whirlwinds, squall lines, are as a rule associated with high values of helicity bulk density $\chi = \mathbf{v} \cdot \nabla \times \mathbf{v}$. This quantity has the dimension of acceleration, so it is convenient to be measured in units of gravity acceleration, g. In intense atmospheric vortices, $|\chi| \sim g$ by the order of magnitude. An appropriate helicity index is the so-called relative helicity

$$\sigma = \frac{\vec{v} \cdot \nabla \times \vec{v}}{|\nabla \times \vec{v}||\vec{v}|} = \cos\left(\widehat{\vec{v}\nabla \times \vec{v}}\right),$$

i.e., the cosine of the angle between the velocity and vorticity vectors. As a rule, in well-developed mesoscale vortices $|\sigma| \approx 1/2$. The maximum possible value $|\sigma| = 1$ is attained in the so called Beltrami flows when $(\nabla \times v) \times v = 0$ or, more generally, when $\vec{\omega}_a \times v = 0$. Here, $\vec{\omega}_a = \lambda v$, the parameter λ being an arbitrary function of spatial coordinates and time. In a stationary Beltrami flow not subjected to diabatic heating, the Ertel's potential vorticity vanishes identically

$$I = \frac{\vec{\omega}_a \cdot \nabla s}{\rho} = \lambda \frac{v \cdot \nabla s}{\rho} = 0.$$

In aerodynamics, the Beltrami flow concept is used to describe the genesis of vortices occurring at the edge of an aircraft wing (Loitsyansky, 1973). In meteorology, under definite limitations, this concept could be applied in order to describe the structure of intense atmospheric vortices (see, e.g., Lilly, 1982; Slezkin, 1990; Bluestein, 1992). To understand the origin of these vortices better, more detailed further analysis of both observational data and numerical and laboratory modeling results is needed not only from the standpoint of such customarily utilized characteristics as energy and vorticity, but also from the viewpoint of helicity.

1.5 Energy and entropy of the atmosphere

The Earth's atmosphere is found to be in a ceaseless motion. To support it against dissipation a permanent feed of Solar radiation is needed. This energy supply is measured in terms of the solar constant $f = 1,370 \pm 10$ $W \cdot m^{-2}$. The Earth's atmosphere absorbs only that fraction of the total energy radiated by the Sun which is screened by the area πa^2, a being the Earth's radius. Due to the diurnal Earth's rotation this radiation is diluted over the total Earth's area $4\pi a^2$, so that the net solar radiation flux is equal to $0.25f \cong 350 \ W \cdot m^{-2}$. With the account for the Earth's planetary albedo A $\cong 0.3$, one arrives at a more accurate estimate of $0.25(1 - A)f \cong 250 \ W \cdot m^{-2}$. As a characteristic time rate of kinetic energy dissipation into heat, we adopt the estimate $D = 5 \ W \cdot m^{-2}$ (Brunt, 1941). Thus, the efficiency of the atmosphere as a heat engine is close to 2%. If, as recommended by Lorenz (1967), we use a later estimation of $D = 2.3$ W·m^{-2} by Oort (1964), we arrive at an approximately two times smaller efficiency of ~1%. It is interesting to support such a 'naive' approach by a more thorough consideration of atmospheric energetics, invoking the general principles of thermodynamics.

The energy balance equation

$$\frac{\partial}{\partial t}\left\{\rho\left(\frac{\mathbf{v}^2}{2}+e+\Phi\right)\right\}+\nabla\cdot\left\{\rho\mathbf{v}\left(\frac{\mathbf{v}^2}{2}+h+\Phi\right)\right\}=0. \tag{1}$$

is a direct consequence of the governing equations of an inviscid compressible fluid subjected to potential external forces only, when there is no diabatic heating. Here, $\rho\mathbf{v}^2/2$ is the kinetic energy bulk density, ρe is the internal energy bulk density, and ρh is the enthalpy bulk density. The gravity field, with potential Φ, is taken into account, so $\rho\Phi$ is the potential energy bulk density. As the Coriolis force is gyroscopic, Equation (1) keeps its form both in absolute and rotating frames of reference, with the accuracy of Φ to be replaced by $\tilde{\Phi}=\Phi-\left(\vec{\Omega}\times\mathbf{x}\right)^2\big/2$. When both viscous friction and diabatic heating are present, the energy equation becomes

$$\frac{\partial}{\partial t}\left\{\rho\left(\frac{\mathbf{v}^2}{2}+e+\tilde{\Phi}\right)\right\}+\nabla\cdot\left\{\rho\mathbf{v}\left(\frac{\mathbf{v}^2}{2}+h+\tilde{\Phi}\right)-\mathbf{I}-\lambda\nabla T-\mathbf{S}\right\}=0.$$

Vector $-\mathbf{I}$ describes the energy flux due to viscosity and has the components $-I_k = -v_i\sigma_{ik}$, where

$$\sigma_{ik}=\eta\left(\frac{\partial v_i}{\partial x_k}+\frac{\partial v_k}{\partial x_i}-\frac{2}{3}\frac{\partial v_m}{\partial x_m}\delta_{ik}\right)+\zeta\frac{\partial v_m}{\partial x_m}\delta_{ik}$$

is the tensor of viscous stresses. Here, tensorial notations are used and the repeated indices denote the summing up. This equation incorporates both the heat fluxes due to temperature gradients $-\lambda\nabla T$, where λ is the thermal conductivity, and the fluxes $-\mathbf{S}$ of radiation being absorbed and re-radiated by atmospheric gases and aerosol. The divergent form of the energy equation excludes the existence of its internal sources or sinks. The air is assumed to be efficiently dry, i.e., latent heating is disregarded, although it could be essential at the lowest tropospheric levels and especially in the equatorial troposphere.

Consideration of atmospheric energetics is associated with definite specifics. Compressible air is subjected to gravitational forces, with the gravity acceleration $g = |\nabla\Phi|$ being of significant magnitude. As the result, large-scale atmospheric circulation systems are quasi-static $(-\nabla p - \rho\nabla\tilde{\Phi} = 0)$, and the atmosphere is stably stratified in the vertical plane. Thus, the preferred direction exists which coincides with that of the action of the gravity force, and fluid dynamic equations lose their primary symmetry. In the frame of a quasi-static approximation, for a perfect gas without viscosity and diabatic heating, the energy equation takes the form (Brunt, 1934; Van Mieghem, 1973)

$$\frac{\partial}{\partial t}\left\{\rho\left(\frac{\mathbf{u}^2}{2} + c_v T + gz\right)\right\} + \nabla_h \cdot \left\{\rho\mathbf{u}\left(\frac{\mathbf{u}^2}{2} + c_v T + gz + \frac{p}{\rho}\right)\right\}$$

$$+ \frac{\partial}{\partial z}\left\{\rho w\left(\frac{\mathbf{u}^2}{2} + c_v T + gz + \frac{p}{\rho}\right)\right\} = 0. \tag{2}$$

Here, $\mathbf{u}$ is the horizontal component of air parcel velocity $\mathbf{v}$. This is just what is called the wind in meteorology. Further on, ∇_h is the horizontal symbolic Hamilton's operator, and z is altitude above the sea level. In large-scale atmospheric circulation systems the vertical velocity w is at least two orders of magnitude smaller than $|\mathbf{v}|$ (in fact, at mid and high latitudes the former is by three orders of magnitude smaller than the latter). Thus, Equation (2) is a very good approximation to Equation (1). After integrating Equation (2) over the entire atmospheric volume, we arrive at the conservation law for total atmospheric energy

$$\frac{d}{dt}\iiint\left\{\frac{\mathbf{u}^2}{2} + c_v T + gz\right\}\rho d\tau = 0. \tag{3}$$

Under a quasi-static approximation, for a dry air and in the absence of mountains

$$\iiint gz\rho d\tau \equiv \iiint RT\rho d\tau,$$

and potential and internal energy terms in Equation (3) cannot change in time independently but are rigidly linked by the ratio R/c_v. This statement is also valid for individual air columns of infinite vertical extent. So, the sum of potential and internal energy $\iiint c_p T\rho d\tau$ deserves to be considered as a unified form of energy called the total potential energy (Margules theorem). Actually, this is the total atmospheric enthalpy. Due to what has been said above, the conservation law (3) can be re-written in the form

$$\frac{d}{dt} \iiint \left\{ \frac{\mathbf{u}^2}{2} + c_p T \right\} \rho d\tau = 0. \tag{4}$$

Equation (4) is the basis for further discussion in this Section.

Still, the analysis of atmospheric energetics directly in terms of kinetic energy (KE) and total potential energy (TPE) is not advantageous by at least two reasons. First, these two components of the total energy are incomparable in magnitude, KE being of the order of fractions of one percent of TPE. Second, when using the KE and TPE concepts directly, it is impossible to gain a clear understanding of energy conversion in the atmosphere. TPE converts into KE of large-scale atmospheric eddies at the expense of the instantaneously occurring sloping thermal convection in the atmosphere (ascent and simultaneous poleward displacement of warm air parcels; descent and equatorward translations of cool ones). Just the same amount of KE is dissipated by viscosity and goes for the increase of TPE. This problem is discussed in more detail in Lorenz (1967) and Van Mieghem (1973). That is why the concept of available potential energy (APE) has become of great importance for meteorology, and its use successfully eliminated both these two difficulties. Nowadays, there are, at least, two main approaches to the estimation of APE. The first approach proposed by Lorenz (1955) is based on the construction of a mechanically stable atmospheric state which corresponds to a minimum APE value. This state is called the reference state. It is constructed by the adiabatic rearrangement of air masses with the conservation of the total air mass within a layer enclosed between two arbitrary isentropic surfaces. The reference atmospheric state is barotropic, the air pressure being uniformly distributed along the isentropic surfaces. From this it follows, in particular, that the reference state is a state of comparison only but not a really attainable atmospheric state. Available potential energy is determined by the difference between the TPE values of the actual and the respective reference states. Calculations show that KE and APE become comparable in magnitude and, what is the most important, only a small fraction of KE converts into APE via viscous dissipation. In other words, because of viscous dissipation the sum of KE and APE decreases, i.e., behaves as if it were the total atmospheric entropy taken with the minus sign (Lorenz, 1967).

The second approach to the estimation of APE, more general from the standpoint of thermodynamics, is based on the construction of a reference state which is stable not only mechanically but also thermodynamically. This reference state has the same value of total entropy as the actual atmospheric state, and the physical process leading to it could be treated as a sequence of idealized thermodynamical Carnot cycles. The difference between the TPE values of actual and respective thermodynamically equilibrium reference states can be called the 'atmospheric free energy'. This concept was first introduced by Obukhov (1949) as a measure of temperature inhomogeneity in an incompressible fluid turbulent flow. It is clear that the free energy always exceeds the Lorenz APE by its magnitude. In fact, even in the state with the vanishing value of the latter the atmosphere has a definite amount of free energy although it could not be directly converted into kinetic energy. This can be done indirectly, e.g., with the help of a thermocouple connected to an electric motor. That is why the free energy concept overestimates the kinetic energy production as compared with the Lorenz APE concept. On the other side, the resulting estimate is thermodynamic. It means that the higher degree of kinetic energy generation than it is predicted by the free energy theory is impossible because in the opposite case it would contradict the general thermodynamic principles.

Various aspects of the available potential energy theory, including those to be discussed in the context of general thermodynamic statements, are highlighted in Marquet (1991); a detailed historical sketch of the problem is given starting from the works of the founders of thermodynamics: Kelvin, Maxwell and Gibbs (see also pp. 33–54 of Gibbs' Collected Works (1928)).

The general approach to the problem of available potential energy is simpler than that by Lorenz and we shall start with it. An account for the rigorous formulation of the available potential concept after Lorenz, which needs the usage of a special isentropic coordinate system, is reserved for Chapter 4.

A starting point is the equation of the second law of thermodynamics written for the case of a perfect gas

$$T\frac{Ds}{Dt} = c_p\frac{DT}{Dt} - \frac{RT}{p}\frac{Dp}{Dt}.$$

We divide both sides of the equation by temperature T and integrate it over the entire atmospheric volume using a well-known Leibnitz' rule for the differentiation of integrals

$$\frac{d}{dt}\iiint s\rho d\tau = \frac{d}{dt}\iiint c_p \ln T\rho d\tau - \frac{d}{dt}\iiint R \ln p\rho d\tau.$$

We assume that after a finite time-interval τ_0 the atmosphere reaches a certain isothermal state reversibly, i.e., with the total entropy $S = \iiint s\rho d\tau$ preserved. During this transition, the heat exchange between air parcels constituting the atmosphere and surrounding bodies (the Sun, the solid Earth) is permitted. The only requirement is that after the interval τ_0 the S value must coincide with that for the initial state. It is additionally assumed that in the final isothermal state the air pressure is uniform over the Earth's surface, when mountains are not considered. In the presence of orography, the surface air pressure varies along with the orography height according to the barometric formula. Denoting all the variables corresponding to the final state by (*), we have

$$\iiint c_p \ln T \rho \, d\tau - \iiint c_p \ln T^* \rho^* \, d\tau$$

$$- \iiint R \ln p \rho \, d\tau + \iiint R \ln p^* \rho^* \, d\tau = 0.$$

We assume that the initial state is quasi-static (the final state is by all means hydrostatic), and transform two last integrals correspondingly as

$$\iiint c_p \ln T \rho \, d\tau - \iiint c_p \ln T^* \rho^* \, d\tau - \iiint R g^{-1} \ln p \, dp \, d\sigma$$

$$+ \iiint R g^{-1} \ln p^* \, dp^* \, d\sigma = 0.$$

Using the atmospheric mass constancy $\iiint \rho \, d\tau = \iiint \rho^* \, d\tau$, we calculate the integrals in question denoting the surface pressure by p_0:

$$\iiint c_p \ln T \rho \, d\tau - c_p \ln T^* \iiint \rho^* \, d\tau$$

$$- \iint_\Sigma R g^{-1} p_0 (\ln p_0 - 1) \, d\sigma + \iint_\Sigma R g^{-1} p_0^* (\ln p_0^* - 1) \, d\sigma = 0, \quad (5)$$

where integration in double integrals is extended over the Earth's surface Σ.

We introduce the averaging operators both over the entire atmospheric volume and the Earth's surface, respectively:

$$\overline{A} = \iiint A \rho \, d\tau \Big/ \iiint \rho \, d\tau, \quad \overline{\overline{B}} = \iint_\Sigma B \, d\sigma \Big/ \iint_\Sigma d\sigma.$$

Note that $p_0^* = \overline{\overline{p}}_0$, because the atmospheric mass is constant. From Equation (5), it follows that

$$\ln T^* = \overline{\ln T} - \frac{R}{c_p}\overline{\left(p_0 \ln p_0 - p_0^* \ln p_0^*\right)}\Big/\overline{p_0}.$$

According to Margules theorem, the total potential energy of the atmosphere over the Earth's relief of height $z = h$, where h is a function of both latitude and longitude, is equal to

$$P = \iiint c_p T\rho d\tau + \iint_\Sigma p_0 h d\sigma = c_p \overline{T}\iiint \rho d\tau + \overline{p_0 h}\iint_\Sigma d\sigma.$$

Without the loss of generality it is assumed that $\overline{\overline{h}} = 0$. In the final state

$$P^* = c_p T^*\iiint \rho d\tau + \overline{p_0^* h}\iint_\Sigma d\sigma,$$

and as the result, the free atmospheric energy is given by the formula (here, the condition of atmospheric mass constancy $p_0^* = \overline{\overline{p}}_0$ is used once again)

$$F = P - P^* = c_p\left(\overline{T} - T^*\right)\iiint \rho d\tau + \overline{\left(p_0 h - p_0^* h\right)}\iint_\Sigma d\sigma$$

$$= c_p\left\{\overline{T} - \exp\left\{\overline{\ln T} - \frac{R}{c_p}\overline{\left(p_0 \ln p_0 - p_0^* \ln p_0^*\right)}\Big/\overline{p_0}\right\}\right\}\iiint \rho d\tau$$

$$+ \overline{\left(p_0 h - p_0^* h\right)}\frac{g}{\overline{\overline{p_0}}}\iiint \rho d\tau$$

$$= c_p\left\{\overline{T} - \exp\{\overline{\ln T}\}\exp\left\{-\frac{R}{c_p}\overline{\left(p_0 \ln p_0 - p_0^* \ln p_0^*\right)}\Big/\overline{p_0}\right\}\right.$$

$$+ \frac{g}{c_p \overline{\overline{p_0}}}\overline{\left(p_0 h - p_0^* h\right)}\left.\right\}\iiint \rho d\tau.$$

Taking T and p_0 in the form $T = \overline{T} + T'$, $p_0 = \overline{\overline{p}}_0 + p_0''$ and using a linearized version of the barometric formula $p_0^* \approx \overline{\overline{p}}_0(1 - gh/R\overline{T})$ valid with the accuracy of the terms of the second order of magnitude inclusively, we find that

$$F \approx \frac{1}{2} c_p \overline{T} \left\{ \overline{\left(T'/\overline{T} \right)^2} + \frac{R}{c_p} \overline{\left(\frac{p_0''}{p_0} + \frac{gh}{R\overline{T}} \right)^2} \right\} \iiint \rho \, d\tau$$

$$= \iiint \frac{1}{2} c_p \overline{T} \left(\frac{T'}{\overline{T}} \right)^2 \rho \, d\tau + \iint_\Sigma \frac{1}{2} \frac{R\overline{T}}{g} \overline{p_0} \overline{\left(\frac{p_0''}{p_0} + \frac{gh}{R\overline{T}} \right)^2} \, d\sigma. \qquad (6)$$

Atmospheric free energy appears as the sum of two terms: (i) three-dimensional energy which originates from the consideration of isobaric processes (Obukhov, 1949), and (ii) two-dimensional energy which characterizes the atmospheric air mass rearrangement in the course of reconstruction towards the thermodynamically equilibrium state.

According to Pearce (1978), who introduced the concept of available potential energy in a somewhat different way (under quadratic approximation it coincides with the first term in Equation (6), see below in more detail), we adopt that

$$\left| T'/\overline{T} \right|_{max} = 0.25, \quad \overline{\left(T'/\overline{T} \right)^2} = 0.02,$$

and $\overline{T} = 253$ K. Setting $\overline{\overline{p}}_0 = 1013$ hPa and assuming the absence of mountains, we take $(72.0 \text{ hPa}^2)^{1/2} = 8.5$ hPa as the characteristic scale of surface air pressure deviation from the mean value (Dobryshman et al., 1982). As the result, the ratio of the surface (without the account for orography) and volume integrals is characterized by a factor of the order of 10^{-3}. Thus, when approximating the free energy in Equation (6) by its first term, we obtain $\overline{F} = 25.4 \times 10^2 \text{ m}^2 \cdot \text{s}^{-2}$, which is roughly 20 times as large as the characteristic value of specific kinetic energy per unit mass, $\overline{K} = 1.3 \times 10^2 \text{ m}^2 \cdot \text{s}^{-2}$ (Oort, 1964).

The ratio $\varepsilon = F/P = 0.5 \left[\overline{\left(T'/\overline{T} \right)^2} \right]$ specifies the fraction of the total potential energy (total enthalpy) which could be spent on the generation of the kinetic energy of atmospheric motions. According to the above data, $\varepsilon \approx 1\%$. This is consistent with the estimate of the atmospheric heat engine efficiency derived earlier by means of the power arguments.

Now, we shall evaluate the increase in the total atmospheric entropy S

after the complete spatial uniformization of both temperature and surface pressure fields (in the presence of mountains, pressure is allowed to vary along with the mountain height according to the barometric formula) provided the total energy

$$E = \iiint \left\{ \frac{1}{2} \mathbf{u}^2 + c_p T \right\} \rho \, d\tau + \iint_\Sigma p_0 h \, d\sigma$$

and the total atmospheric mass are constant. It could be written (all the variables corresponding to the final state are marked by (**)) that

$$\Delta S = S^{**} - S = \iiint c_p \ln T^{**} \rho^{**} d\tau - \iiint c_p \ln T \rho d\tau$$

$$+ \iint_\Sigma Rg^{-1} p_0 \ln p_0 d\sigma - \iint_\Sigma Rg^{-1} p_0^{**} \ln p_0^{**} d\sigma$$

$$= \frac{1}{T^{**}} \iiint c_p \left(T^{**} - T \right) \rho d\tau$$

$$+ \iiint c_p \left(\ln T^{**} - 1 - \ln T + \frac{T}{T^{**}} \right) \rho \, d\tau$$

$$+ \iint_\Sigma Rg^{-1} \left(p_0 \ln p_0 - p_0^{**} \ln p_0^{**} \right) d\sigma,$$

and, besides, $\bar{\bar{p}}_0 = \bar{\bar{p}}_0^{**}$, because the atmospheric mass is constant. As a consequence of energy conservation law, we have

$$\iiint c_p \left(T^{**} - T \right) \rho d\tau = \iiint \frac{1}{2} \mathbf{u}^2 \rho d\tau + \iint_\Sigma \left(p_0 - p_0^{**} \right) h \, d\sigma$$

and, thus,

$$\Delta S = \frac{1}{T^{**}} \left\{ \iiint \frac{1}{2} \mathbf{u}^2 \rho d\tau + \iint_\Sigma \left(p_0 h - p_0^{**} h \right) d\sigma \right.$$

$$+ \iiint c_p T^{**} \left(\ln T^{**} - 1 - \ln T + \frac{T}{T^{**}} \right) \rho \, d\tau$$

$$+\iint_{\Sigma} \frac{RT^{**}}{g}\left(p_0 \ln p_0 - p_0^{**} \ln p_0^{**}\right)d\sigma\Bigg\} \approx \frac{1}{T^{**}}\Bigg\{\iiint \frac{1}{2}\mathbf{u}^2 \rho d\tau$$

$$+\iiint \frac{1}{2}c_p T^{**}\left(\frac{T-T^{**}}{T^{**}}\right)^2 \rho d\tau + \iint_{\Sigma} \frac{1}{2}\frac{RT^{**}}{g}\overline{\overline{p}}_0\left(\frac{p_0-\overline{\overline{p}}_0}{\overline{\overline{p}}_0} + \frac{gh}{RT^{**}}\right)^2 d\sigma\Bigg\}, \quad (7)$$

if the terms squared in $(T-T^{**})/T^{**}$ and $(p_0 - \overline{\overline{p}}_0)/\overline{\overline{p}}_0$ are retained and the linear version of the barometric formula $p_0^{**} \cong \overline{\overline{p}}_0(1 - gh/RT^{**})$ is used.

The problem of the relation between temperatures T^{**} and $\overline{T}$ is easily solved in the case without orography. Here, one has

$$\iiint c_p\left(T^{**} - T\right)\rho d\tau = \iiint \frac{1}{2}\mathbf{u}^2 \rho d\tau.$$

On the other hand, by definition, $\iiint c_p\left(\overline{T} - T\right)\rho d\tau = 0$ and $T^{**} = \overline{T} + 0.5\overline{\left(\mathbf{u}^2\right)}/c_p$, i.e., T^{**} exceeds $\overline{T}$ by the value of 10^{-3} in relative units. The temperatures T^{**} and $\overline{T}$ are close also in the presence of mountains, though it is not easy to determine the sign of inequality: $T^{**} < \overline{T}$ or $T^{**} > \overline{T}$. When replacing T^{**} by $\overline{T}$ in Equation (7), we obtain

$$\Delta S \approx \frac{1}{\overline{T}}(K + F), \quad (8)$$

where K is the kinetic energy of atmospheric motions. Under the influence of irreversible factors (viscosity, thermoconductivity, radiation exchange) the energy sum $K + F$ monotonously decreases along with the entropy deficit ΔS, the latter quantity being an integral parameter which characterizes the deviation between the actual atmospheric state and the ultimate state of complete thermodynamic equilibrium (Dutton, 1973).

Further on, we shall discuss briefly Pearce's (1978) approach to the available potential energy problem. The starting point is to use both the entropy equation

$$\frac{d}{dt}\overline{s} = \overline{\left(\hat{Q}/T\right)}, \quad \overline{s} = c_p \overline{\ln T}$$

and the equation for the total atmospheric potential energy (without account for mountains)

$$\frac{d}{dt}\overline{P} = \overline{\hat{Q}} - \overline{C}, \quad \overline{P} = c_p \overline{T}.$$

Here, C is the rate of potential-to-kinetic energy conversion. A constant quantity Θ, with the dimension of temperature, is introduced and the linear combination

$$\frac{d}{dt}\left(\overline{P} - \Theta\overline{s}\right) = \overline{\hat{Q}\left(1 - \Theta/T\right)} - \overline{C}$$

is constructed. Further on, this equation is integrated over time under assumption that starting from an actual atmospheric state it is possible to arrive at an isothermal atmospheric state with temperature Θ, after a finite time-interval τ_1

$$\overline{P}\left(\tau_1\right) - \overline{P}(0) - \Theta\overline{s}\left(\tau_1\right) + \Theta\overline{s}(0)$$

$$= \int_0^{\tau_1}\overline{\hat{Q}\left(1 - \Theta/T\right)}dt - \int_0^{\tau_1}\overline{C}dt. \tag{9}$$

The constant Θ is chosen in such a way that $\Theta^{-1} = \left(\overline{T^{-1}}\right)$, i.e.,

$$\Theta = \left[\left(\overline{T^{-1}}\right)\right]^{-1} = \overline{T}\left(1 - \overline{\left(T'/\overline{T}\right)^2} + ...\right)$$

and, consequently, $\Theta < \overline{T}$. The very idea of the choice of this particular Θ is the resulting extreme simplicity of the first integrand in the right-hand side of Equation (9)

$$\overline{\hat{Q}\left(1 - \Theta/T\right)} = \overline{\left(\hat{\overline{Q}} + \hat{Q}'\right)\left[\left(\overline{T^{-1}}\right) - T^{-1}\right]}\Theta = -\overline{\hat{Q}'\Theta T^{-1}},$$

where only $\hat{Q}'$, i.e., the deviation of the heating term from its mass-averaged value, plays the role. Thus, the net contribution to APE generation is only due to diabatic heating unevenly distributed in space. From this point of view, the dissipative processes create a quasi-uniform heating

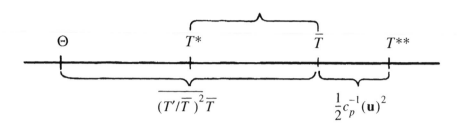

FIGURE 1 Relation between characteristic temperatures Θ, T^*, $\overline{T}$ and T^{**} in atmospheric free energy theory.

background in space and, in fact, do not contribute to APE generation. Along with this, the total atmospheric enthalpy, determined by the mean temperature $\overline{T}$, increases. A factor, which sets an upper limit to $\overline{T}$ growth, is the outgoing infrared radiation of the atmosphere. The corresponding radiative equilibrium temperature T_e is determined through the equation $0.25 f (1 - A) = \sigma T_e^4$ (σ is the Stefan–Boltzmann constant) and is equal to 255 K (see, e.g., Golitsyn, 1973), which is close to $\overline{T} = 253$ K. It is worth mentioning that on other planets of the Solar system (on Venus and Mars) the temperatures $\overline{T}$ and T_e are not close at all. To explain this, special argumentation is necessary which is beyond the scope of this book.

The left-hand side of Equation (9) takes the form

$$c_p(\overline{T} - \Theta) - c_p \Theta \overline{\ln T} + c_p \Theta \ln \Theta$$

$$= c_p \overline{T} \overline{(T'/\overline{T})^2} - c_p \overline{T} \left\{ \ln \overline{T} - \frac{1}{2} \overline{(T'/\overline{T})^2} - \ln \overline{T} + \overline{(T'/\overline{T})^2} \right\} + \ldots$$

$$= \frac{1}{2} c_p \overline{T} \overline{(T'/\overline{T})^2} + \ldots,$$

i.e., under quadratic approximation one has the same mathematical expression for the free energy as in the frame of approach developed before. The term

$$-\overline{\hat{Q}'\Theta T^{-1}} \approx -\overline{\hat{Q}'\overline{T}T^{-1}} \approx \left(\overline{T}\right)^{-1}\overline{\hat{Q}'T'}$$

depicts the time rate of APE generation by diabatic heating. The relation between different characteristic temperatures appearing in the theory is shown in Figure 1. To the accuracy of neglected cubic terms one has $\overline{T} - \Theta = 2(\overline{T} - T^*)$, i.e., $T^* = 0.5(\overline{T} + \Theta)$. According to Pearce (1978), $\Theta = 248$ K and the value of T^*, as we stated before, is between Θ and $\overline{T}$. It was pointed out above that $K \approx 0.05F$, so with good accuracy $\Delta \overline{s} \approx \overline{F}/\overline{T} = 25.4 \times 10^2$ m^{2}s^{-2}/253 K ≈ 10 m^{2}s^{-2}K^{-1}. A natural entropy unit is the value of specific heat at constant pressure $c_p = 1004$ m$^2 \cdot$s$^{-2} \cdot$K^{-1}, so $\Delta \overline{s}/c_p \approx 10^{-2}$. This non-dimensional factor is an indicator of intensity of general atmospheric circulation. The dissipative factors tend to decrease the entropy deficit, i.e., lead the atmosphere to the state of complete thermodynamic equilibrium. Differential heating of the atmosphere by solar radiation increases this deficit and, thus, enables the atmosphere to escape from the ultimate state of 'heat death'.

1.6 Atmospheric angular momentum

Displacements of air masses over the rotating spherical Earth are controlled by one of the most important principles in Newtonian mechanics. It is the angular momentum conservation law. Dealing with this topic, we arrive at the necessity to use the explicit form of atmospheric governing equations written in spherical coordinates. We start from the mathematical expression for the specific kinetic energy of atmospheric motions relative to the rotation of a spherical coordinate system at a constant angular velocity $\bar{\Omega}$ with the poles situated in the Earth's geographic poles and with the origin of coordinates in the solid Earth's center:

$$T = \frac{1}{2}\left(\dot{r}^2 + r^2\dot{\vartheta}^2 + r^2\sin^2\vartheta\left(\dot{\lambda} + \Omega\right)^2\right). \tag{1}$$

Here r is the radius, ϑ is the co-latitude, λ is the longitude and the well-known expression for spherical metrics (the squared distance between two infinitely closely spaced points) is used

$$(dl)^2 = (dr)^2 + r^2(d\vartheta)^2 + r^2\sin^2\vartheta(d\lambda)^2. \tag{1'}$$

The point above the variables in Equation (1) denotes the material time-derivative. The Lagrange equations are further written as (cf. Kochin *et al.*,

1964; Sretensky, 1987)

$$\frac{D}{Dt}\frac{\partial T}{\partial \dot{q}_i} - \frac{\partial T}{\partial q_i} = -\frac{1}{\rho}\frac{\partial p}{\partial q_i} - \frac{\partial \Phi}{\partial q_i} + \frac{1}{\rho}F_i. \tag{2}$$

The spherical coordinates r, ϑ, λ play the role of generalized coordinates q_i ($i=1,2,3$) in classical mechanics, Φ is the gravity potential and $\mathbf{F}$ stands for any non-potential force. Taking into account that the latter is primarily the viscous force, we write it as if it were a bulk force (see Section 1.1). Starting from Equations (1, 2) it is technically not difficult to write down the Euler's equations in rotating spherical coordinates. We need only one of the three resulting equations. Longitude λ is the so-called cyclic coordinate, i.e., it does not enter the spherical metrics (1′) explicitly and the gravitational potential does not depend on longitude, either. Consequently

$$\frac{D}{Dt}\left(r^2\sin^2\vartheta\left(\dot{\lambda}+\Omega\right)\right) = -\frac{1}{\rho}\frac{\partial p}{\partial \lambda} + \frac{1}{\rho}F_\lambda.$$

We arrived at the equation for the axial component of the angular momentum vector. Representing the atmosphere by a thin fluid film of a thickness much less than the Earth's radius a and introducing a zonal component of velocity $u_\lambda = a\sin\vartheta\,\dot{\lambda}$, we obtain the equation for the axial component of the angular momentum in Eulerian variables

$$\frac{DM}{Dt} = -\frac{1}{\rho}\frac{\partial P}{\partial \lambda} + \frac{1}{\rho}F_\lambda, \quad M = \left(u_\lambda + a\Omega\sin\vartheta\right)a\sin\vartheta.$$

We integrate the resulting equation over the entire atmospheric volume

$$\frac{d}{dt}\iiint M\rho\,d\tau = -\iiint \frac{\partial p}{\partial \lambda}d\tau + \iiint F_\lambda d\tau, \tag{3}$$

where $d\tau = a^2\sin\vartheta d\vartheta d\lambda dz$ and $z = r - a$. In the presence of mountains (it is assumed that large-scale orography is described by the equation $z = h(\lambda, \vartheta)$, where h are the distances between the points lying on the Earth's surface and their projections onto a perfect sphere of radius a) the first right-hand side term in Equation (3) is non-zero and equals to

$$-\iint_\Sigma\left\{\frac{\partial}{\partial \lambda}\int_h^\infty p\,dz\right\}d\sigma - \iint_\Sigma p_0\frac{\partial h}{\partial \lambda}d\sigma = -\iint_\Sigma p_0\frac{\partial h}{\partial \lambda}d\sigma.$$

To derive it, the rule of differentiation of integrals with variable integration limits was first used. Secondly, the notation $d\sigma = a^2\sin\vartheta\, d\vartheta\, d\lambda$ is introduced for the area of an element of the Earth's surface Σ, and $p_0 = p(\lambda,\vartheta,h(\lambda,\vartheta),t)$ stands for surface air pressure. Equation (3) can be re-written in a more symmetric form

$$\frac{d}{dt}\iiint M\rho\, d\tau = -\frac{1}{2}\iint_\Sigma \left(p_0\frac{\partial h}{\partial\lambda} - h\frac{\partial p_0}{\partial\lambda}\right) d\sigma + \iiint F_\lambda\, d\tau,$$

where both the surface pressure p_0 and the mountain height h enter in equivalent terms.

Variations of atmospheric angular momentum are caused, first, by the orographic torque due to the differences between pressure p_W and p_E on the western and eastern mountain slopes taken at the same altitude and latitude. In particular, in the lee of meridionally oriented mountain ridges (such as Rocky Mountains and Andes) overblown by westerlies, one usually observes pressure troughs. Thus, the atmospheric wind 'pushes' mountains, and also the entire solid Earth, in the eastward direction. In turn, mountains push the air in the opposite, westward direction, i.e., damp the westerlies. Secondly, the variations of atmospheric angular momentum are caused by the surface friction torque.

Let us introduce a tensor of frictional stresses T_{ik} ($i,k = 1, 2, 3$) such that $F_i = \partial T_{ik}/\partial x_k$, by definition. Using Gauss–Ostrogradsky's theorem, one has

$$\frac{d}{dt}\iiint M\rho\, d\tau = -\frac{1}{2}\iint_\Sigma \left(p_0\frac{\partial h}{\partial\lambda} - h\frac{\partial p_0}{\partial\lambda}\right) d\sigma + \iint_\Sigma T_{ik}n_k\, d\sigma, \quad i = 3, \quad (4)$$

where the unit vector $\mathbf{n}$ is orthogonal to the Earth's surface and is oriented downward, beneath the ground. It is assumed that frictional stresses vanish at the top of the atmosphere. The second integral in the right-hand side of Equation (4) can be approximately taken at $z = 0$ and is used in the form $-\iint T_\lambda\, d\sigma$, where T_λ is the zonal component of frictional stresses on the Earth's surface. It is usually assumed that

$$T_\lambda = c_D\rho\left(u_\lambda^2 + u_\vartheta^2\right)^{1/2} u_\lambda\, a\sin\vartheta,$$

where the numerical coefficient c_D is determined on the basis of both empirical and experimental data processing and is of the order of 10^{-3}. The values of c_D depend on both the stratification of the atmospheric boundary layer and underlying physical surface properties. For example, $c_D = 0.0013$ over oceans (see Lorenz, 1967). In more detail, the dissipative processes in the atmospheric boundary layer are discussed in Chapter 5.

In the presence of mountains and under the action of surface friction, we can speak of the conservation of the angular momentum for the entire 'atmosphere – solid Earth' system

$$\iiint M\rho\, d\tau + M_\oplus = \text{invariant}.$$

Here, $M_\oplus = I_\oplus \Omega$ is the axial component of the solid Earth angular momentum, and $I_\oplus = 7.04 \times 10^{37}$ kg·m^2 is the corresponding principal momentum of inertia (Barnes et al., 1983). Any variation of the atmospheric angular momentum results in a change of the solid Earth angular momentum, equal in magnitude but of opposite sign

$$\delta \iiint M\rho\, d\tau = -\delta M_\oplus = -I_\oplus \delta\Omega.$$

Currently, the length-of-day (l.o.d.) $\Lambda = 2\pi\Omega^{-1}$, averaged over 5 days, is reliably measured by precise astronomical methods (see, e.g., Bureau International de l'Heure, 1979, 1980). Using the notation Δ^* for l.o.d. Λ deviations from the standard value $\Lambda_0 = 86400$ s, one has, under a linear approximation

$$\delta \iiint M\rho\, d\tau = I_\oplus \frac{2\pi\, \delta\Delta^*}{\Lambda_0^2}.$$

Because the net atmospheric mass transport across latitudinal circles is small, the atmospheric angular momentum variations are mainly caused by the fluctuations of the relative angular momentum

$$M = \iiint u_\lambda a \sin\vartheta\, \rho\, d\tau$$

determined by the zonal wind variability

$$\delta \iiint M\rho\, d\tau \approx \delta M.$$

Terrestrial atmosphere is observed in a state of super-rotation when air parcels in their zonal motion (along latitudinal circles) leave the Earth behind themselves. This occurs due to atmospheric baroclinity. Air temperature decreases towards the poles in the main atmospheric bulk but air pressure varies in the meridional direction in much smaller extent. Resulting slopes between isothermal and isobaric surfaces can persist only due to the action of the Coriolis force. This implies an increase of the positive (eastward-directed) wind component u_λ with altitude (which is called the thermal wind) which, because surface winds are rather weak due to viscous friction, leads to the positive values of the relative atmospheric angular momentum with the dominant contribution of the thermal wind. According to Sidorenkov (1976), the estimate of the relative atmospheric angular momentum is 12.8×10^{25} kg m$^2 \cdot$s^{-1}. When divided by the total atmospheric mass $m_A = 5.3 \times 10^{18}$ kg, it gives $\overline{M} = M/m_A = 2.4 \times 10^7$ m^{2}s^{-1}. If we imagine that the atmosphere rotates like a rigid body and has just this value of relative angular momentum, it would correspond to the equatorial air parcel linear velocity $U = 1.5 \left(\overline{M}/a \right)^2 = 5.7$ m·s^{-1}. Thus, roughly speaking, individual air parcels, leaving the Earth behind themselves, make one revolution around the Earth per approximately 80 days.

Based on these estimates, let us calculate the decrease of l.o.d. in an hypothetical case when large-scale air motion in the atmosphere ceases and the solid Earth gains all the angular momentum the atmosphere had before. Here, $\delta\Lambda^* = 21.6 \times 10^{-4}$ s $\cong$ 2 ms. It means that the length of year would decrease by less than 1 s. With relative accuracy approaching 10^{-8} $(10^{-6}\%)$ for motions studied in dynamic meteorology, the Earth's rotation angular velocity can be considered as a constant, and the solid Earth can be regarded as an angular momentum reservoir of infinite capacity, thus invoking an analogy with thermodynamics where a thermostat is treated as a heat reservoir of infinite capacity.

A more complete theory should take into account the vectorial essence of the concept of the atmospheric angular momentum

$$\vec{M}_A = \iiint \left\{ \mathbf{x} \times \mathbf{v} + \mathbf{x} \times \left(\vec{\Omega} \times \mathbf{x} \right) \right\} \rho \, d\tau,$$

i.e., the existence of two atmospheric angular momentum equatorial components, in addition to the axial component (see Figure 2). The consequences of the exchange of angular momentum between the atmosphere and the solid Earth are not only l.o.d. changes but also excursions of the geographic poles along the Earth's surface. The observed displacements of the rotation axis occurring within the limit of several meters correspond to relative variations in the magnitude of the equatorial components of $\vec{M}_A$, of the order of 10^{-6}. Account of these problems is given in detailed papers by R. Hide and co-authors (Hide et al., 1980; Barnes et al., 1983; Bell et al.,

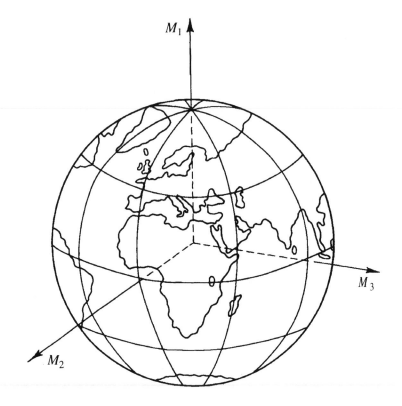

FIGURE 2 Axial M_1 and equatorial M_2, M_3 components of the atmospheric angular momentum vector. When the atmosphere is represented by a thin film of incompressible fluid, a rigid-body-rotation about axis 1 is described by a stream function proportional to $P_1^0(\cos\vartheta)$, a rigid-body-rotation about axis 2 by $P_1^1(\cos\vartheta)\cos\lambda$ ($\lambda = 0$ corresponds to the Greenwich meridian) and about axis 3 by $P_1^1(\cos\vartheta)\sin\lambda$.

1991) which give impressive examples of intercomparison between the coordinates of the instantaneous Earth's rotation axis computed with the help of atmospheric General Circulation Models and the astronomically observed coordinates. In these papers a reader will find a vast bibliography on the subject.[6]

Treatment of the problem of temporal variations in the value of atmospheric angular momentum is related directly to the fundamental aspects of atmospheric energetics. In the absence of kinetic energy generation the atmosphere would rotate together with the solid Earth like a rigid body with a vanishing relative angular velocity. It would be a state of the absolute

---

[6] More recently, the role of the equatorial components of the atmospheric angular momentum was investigated in Bell (1994) and Egger and Hoinka (1999).

kinetic energy minimum of the 'atmosphere – solid Earth' system, provided the total angular momentum of this coupled system is a prescribed constant. Differential heating of the terrestrial atmosphere by the Sun drives the atmospheric heat engine. Due to this, the angular momentum is redistributed between the atmosphere and the solid Earth and the former gains a positive relative angular momentum (due to westerlies). The concept of angular momentum can be directly incorporated into atmospheric energetics via the concept of unavailable kinetic energy. This is the kinetic energy of atmospheric rigid-body-like superrotation which, under the atmosphere approximation as a thin fluid film, is determined completely by an excessive, if compared to the state of relative rest, atmospheric angular momentum. The latter value is close to the relative atmospheric angular momentum. Specific (per unit mass) unavailable kinetic energy is determined by the formula $K_M = 0.75(\overline{M}/a)^2$ (Kurgansky, 1981) where $\overline{M}$ is the specific relative atmospheric angular momentum. If the intensity of the atmospheric rigid-body-like rotation is measured in terms of the linear velocity of air particles at the equator $U = 1.5(\overline{M}/a)$, then $K_M = (1/3)U^2$. Kurgansky (1981) gave an estimate of K_M for the Northern Hemisphere on the basis of data available at that time. An annual mean K_M value is 6–7% of the total kinetic energy. To date, much more detailed data on the atmospheric angular momentum have appeared (see, e.g., Bell *et al.*, 1991) which, on the whole, confirm the accuracy of Kurgansky's (1981) estimate.

As it was stated above, in order for the atmosphere to keep positive values of the excessive angular momentum it should have a permanent supply of power W. Based on general physical arguments, Sidorenkov (1976) suggested the use of the functional dependence $M \propto W^x$ with $x > 0$. We can try to make the exponent x value concrete based on the concept of unavailable kinetic energy. We estimate its viscous dissipation rate by the formula $W \propto U^3$, which results from the law of quadratic surface friction resistance. Now, it follows that $x = 1/3$. Estimates show that the power required to support the atmospheric superrotation is several percent of the total kinetic energy viscous dissipation rate.

It is necessary to remember that there is a mechanism for the redistribution of angular momentum between the atmosphere and the solid Earth which requires no purposeful power expenses. We mean orography, the influence of which explains the major part of short-term atmospheric angular momentum fluctuations (with periods of few days). Averaging Equation (4) over a sufficiently long time-period, when orographic torque fluctuations due to the atmospheric weather variability are smoothed, we arrive at the requirement of approximate vanishing of the surface friction torque (small systematic terms which account for the angular momentum seasonal course are also to be retained)

$$\iint\limits_{\Sigma} c_D \rho \left(u_\lambda^2 + u_\vartheta^2 \right)^{1/2} u_\lambda a \sin \vartheta d\sigma \approx 0. \tag{5}$$

It is possible to fulfill the requirements of Equation (5) only in the case when the zonal wind u_λ changes its sign somewhere on the Earth's surface. Atmospheric circulation satisfies this necessary condition in the most rational way. Vanishing of u_λ takes place at about 30–35° latitude in both hemispheres. Between these critical latitudes we find the belt of easterlies (trade winds). In mid and high latitudes, we observe more intensive westerlies. The latter are stronger because the distance from the Earth's rotation axis is smaller there. Information on the details of the atmospheric angular momentum budget and a summary of theoretical ideas on the atmospheric general circulation can be found in classical monographs by Starr (1966), Lorenz (1967), Palmen and Newton (1969) and in more recent books by Peixoto and Oort (1992), Wiin-Nielsen and Chen (1993), James (1994). Here, we only emphasize that the budget of angular momentum should be carefully checked in the course of numeric modeling of atmospheric circulation.

The requirements of atmospheric angular momentum conservation law make it necessary to correct some statements in the previous section which discussed problems of atmospheric energetics. In that piece of text, where energetic isolation of the atmosphere is assumed to be necessary, we should, generally speaking, also assume that the atmospheric angular momentum is constant. Indeed, the assumption of the frictionally caused exchange of angular momentum between the atmosphere and the solid Earth contradicts the requirement of atmospheric energy conservation. This is because a certain portion of heat emanated in the course of kinetic energy viscous dissipation is transferred to the solid Earth. Its future fate has to be inspected in a special way taking into account the physical properties of the underlying Earth's surface, e.g., soil thermal conductivity, emissivity, etc. Now, the ultimate state of complete thermodynamical equilibrium becomes a state of rigid-body-like rotation which is fully determined by the value of excessive angular momentum of the actual atmospheric state. As a consequence, in the main formula (8) from Section 1.5 the total kinetic energy should be replaced by the difference $K - K_M$. Following a suggestion by Kurgansky (1981), the latter quantity could be named the available kinetic energy (cf. Starr (1966)). Strictly speaking, the final state of rigid-body-like rotation with non-zero excessive angular momentum is dynamically incompatible with the existence of mountains. A candidate for such a final state could be a stationary solution resulting in the problem of the Earth's relief overflow by an isothermal air stream given the excessive angular momentum. One possible difficulty is the non-uniqueness of such solutions. Hydrodynamic stability requirements could serve as a criterion for the selection of these solutions.

CHAPTER 2

Reduced Equations of Atmospheric Dynamics

This chapter is devoted to the problems of simplification (reduction) of the atmospheric fluid dynamic equations with the account of spatio-temporal scales of the class of motions considered in the monograph. When writing the first three sections, the author followed the ideas of a classical paper 'On the question of geostrophic wind' by A. M. Obukhov (1949a). In Section 2.1 the equations for a two-dimensional atmospheric model are derived, primarily in a more general case than in (Obukhov, 1949a), i.e., taking into account the atmospheric two-dimensional baroclinicity. After that, transition to the case of a barotropic atmosphere is performed. Section 2.2 considers the barotropic atmosphere motion, with infinitesimal deviations from the geostrophic equilibrium state, and the problem of the adjustment of meteorological fields. Particular attention is paid to the energetics of the adjustment process. Section 2.3 reproduces a classical asymptotic procedure of the derivation of nonlinear quasi-geostrophic equations for the case of a barotropic atmosphere. Also, non-linear quasi-geostrophic equations for a baroclinic three-dimensional atmosphere are briefly discussed (giving only efficacious formulae), which are necessary for the following (see Chapters 3 and 5). Section 2.4 considers non-divergent barotropic Rossby waves on a rotating sphere; the author starts from nonlinear equations of motion and treats Rossby wave harmonics as their exact solutions. The fifth section touches upon the relatively recently raised problem of spontaneous emission of waves and the derivation of an evolutionary equation for the slow large-scale component of motion, proceeding from the principle of wave emission minimum. This is of interest in relation to the initialization procedure, which deals with constructing a correct initial atmospheric state for weather forecasting.

2.1 Two-dimensional atmospheric models ('shallow-water' approximation)

Below, we develop a procedure for reducing primitive atmospheric dynamic equations to simplified two-dimensional equations. They are attractive because of their simplicity, clearness, computational cheapness and also their completeness in describing non-linear atmospheric processes.

A system of atmospheric dynamic equations, taken under quasi-static approximation and used in the form of conservation laws, is considered. Thermodynamic processes are assumed to be adiabatic, and external forces to have a potential. For simplicity of notations, Cartesian coordinates are used with the x-axis directed eastward, the y-axis oriented northward, and the z-axis aligned upward; u, v and w are the corresponding velocity components. Thus, we start with the equations:

$$\frac{\partial}{\partial t}(\rho u) + \frac{\partial}{\partial x}(\rho u^2) + \frac{\partial}{\partial y}(\rho u v) + \frac{\partial}{\partial z}(\rho u w) - l\rho v = -\frac{\partial p}{\partial x},$$

$$\frac{\partial}{\partial t}(\rho v) + \frac{\partial}{\partial x}(\rho u v) + \frac{\partial}{\partial y}(\rho v^2) + \frac{\partial}{\partial z}(\rho v w) + l\rho u = -\frac{\partial p}{\partial y},$$

$$-g\rho - \partial p/\partial z = 0,$$

$$\frac{\partial}{\partial t}\rho + \frac{\partial}{\partial x}(\rho u) + \frac{\partial}{\partial y}(\rho v) + \frac{\partial}{\partial z}(\rho w) = 0,$$

$$\frac{\partial}{\partial t}(\rho E) + \frac{\partial}{\partial x}(\rho u E + \rho u R T) + \frac{\partial}{\partial y}(\rho v E + \rho v R T)$$

$$+ \frac{\partial}{\partial z}(\rho w E + \rho w R T) = 0,$$

$$E = \frac{1}{2}(u^2 + v^2) + c_v T + gz, \quad p = \rho R T.$$

Here, l is the doubled vertical component of the Earth's angular velocity Ω (the Coriolis parameter).

The field-averaging operations over height are introduced:

$$\hat{A} = \int\limits_0^\infty A\,dz, \quad \overline{B} = \int\limits_0^\infty B\rho\,dz \Big/ \hat{\rho}\,.$$

The averaged atmospheric dynamic equations take the form:

$$\frac{\partial}{\partial t}\hat{\rho}\overline{u} + \frac{\partial}{\partial x}\hat{\rho}\overline{u^2} + \frac{\partial}{\partial y}\hat{\rho}\overline{uv} - l\hat{\rho}\overline{v} = -\frac{\partial\hat{p}}{\partial x},$$

$$\frac{\partial}{\partial t}\hat{\rho}\overline{v} + \frac{\partial}{\partial x}\hat{\rho}\overline{uv} + \frac{\partial}{\partial y}\hat{\rho}\overline{v^2} + l\hat{\rho}\overline{u} = -\frac{\partial\hat{p}}{\partial y},$$

$$\frac{\partial}{\partial t}\hat{\rho} + \frac{\partial}{\partial x}\hat{\rho}\overline{u} + \frac{\partial}{\partial y}\hat{\rho}\overline{v} = 0,$$

$$\frac{\partial}{\partial t}\hat{\rho}\overline{E} + \frac{\partial}{\partial x}\hat{\rho}\overline{u(E+RT)} + \frac{\partial}{\partial y}\hat{\rho}\overline{v(E+RT)} = 0. \tag{1}$$

The impermeability condition $w = 0$ on the Earth's surface $z = 0$ is used (in the absence of orography) along with the assumption of the lack of vertical fluxes of momentum, mass and energy at the upper boundary of the atmosphere, $z \to \infty$. Let us clarify the meaning of the variables in (1):

$$\hat{\rho} = \int\limits_0^\infty \rho\,dz = -\int\limits_0^\infty g^{-1}(\partial p/\partial z)\,dz = p(x,y,0,t)/g$$

is the two-dimensional air density mass proportional to the surface air pressure;

$$\hat{p} = \int\limits_0^\infty p\,dz = [zp]_0^\infty - \int\limits_0^\infty z(\partial p/\partial z)\,dz = \int\limits_0^\infty gz\rho\,dz$$

is the two-dimensional density of potential energy.

We set $u = \overline{u} + u'$, $v = \overline{v} + v'$, then $\overline{u^2} = \overline{u}^2 + \overline{u'^2}$, $\overline{uv} = \overline{u}\,\overline{v} + \overline{u'v'}$, $\overline{v^2} = \overline{v}^2 + \overline{v'^2}$. It is assumed that the atmospheric wind slightly changes with altitude, i.e., thermal wind is neglected, then one has approximately $\overline{u^2} \approx \overline{u}^2$, $\overline{uv} \approx \overline{u}\,\overline{v}$, $\overline{v^2} \approx \overline{v}^2$. As a result, instead of the first three equations in (1) we will have:

$$\frac{\partial}{\partial t}\hat{\rho}\overline{u} + \frac{\partial}{\partial x}\hat{\rho}\overline{u}^2 + \frac{\partial}{\partial y}\hat{\rho}\overline{uv} - l\hat{\rho}\overline{v} = -\frac{\partial\hat{p}}{\partial x},$$

$$\frac{\partial}{\partial t}\hat{\rho}\overline{v} + \frac{\partial}{\partial x}\hat{\rho}\overline{uv} + \frac{\partial}{\partial y}\hat{\rho}\overline{v}^2 + l\hat{\rho}\overline{u} = -\frac{\partial\hat{p}}{\partial y}, \qquad (2)$$

$$\frac{\partial}{\partial t}\hat{\rho} + \frac{\partial}{\partial x}\hat{\rho}\overline{u} + \frac{\partial}{\partial y}\hat{\rho}\overline{v} = 0.$$

The resulting system is not closed, because the number of unknown variables $\overline{u}, \overline{v}, \hat{\rho}, \hat{p}$ exceeds that of equations by one. To close the system, we use the energy conservation law. When doing it, we apply the identity

$$\int_0^\infty \rho(\mathbf{u} - \overline{\mathbf{u}})(c_p T + gz)dz = \int_0^\infty \left\{ c_p\left(\frac{\partial T}{\partial z} + \frac{g}{c_p}\right)\int_z^\infty \rho(\mathbf{u} - \overline{\mathbf{u}})dz' \right\}dz,$$

with $\mathbf{u} = (u,v)$, which is readily established by integration by parts. Assume that the atmosphere is isentropic in the vertical, i.e., $\partial T/\partial z + g/c_p = 0$, then

$$\int_0^\infty \rho(\mathbf{u} - \overline{\mathbf{u}})(c_p T + gz)dz = 0.$$

Strictly speaking, a neutrally stratified atmosphere has the finite height $H_0 = c_p T_0/g \approx 30$ km, where T_0 is the averaged surface air temperature. Nevertheless, one may disregard this in all formulae and extend integration up to infinitely large altitudes, if one defines the air density field in such a way that $\rho = 0$ at $z > H_0$. According to the Margules' theorem, one has

$$\hat{\rho}\,\overline{E} = \frac{1}{2}\hat{\rho}\overline{(u^2 + v^2)} + \hat{\rho}c_p\overline{T},$$

which, according to the above assumptions, leads to

$$\hat{\rho}\,\overline{E} = \frac{1}{2}\hat{\rho}(\overline{u}^2 + \overline{v}^2) + \hat{\rho}c_p\overline{T}.$$

Further on, it is not difficult to write down that

$$\overline{\hat{\rho}\mathbf{u}(E + RT)} = \overline{\hat{\rho}\mathbf{u}\frac{\mathbf{u}^2}{2}} + \overline{\hat{\rho}\mathbf{u}(c_p T + gz)}$$

$$= \hat{\rho}\mathbf{u}\overline{\frac{\mathbf{u}^2}{2}} + \hat{\rho}\overline{\mathbf{u}}\overline{(c_p T + gz)}$$

$$\approx \hat{\rho}\overline{\mathbf{u}}\frac{\overline{\mathbf{u}}^2}{2} + \hat{\rho}\overline{\mathbf{u}}(c_p + R)\overline{T}$$

As a result, we have

$$\frac{\partial}{\partial t}\left(\hat{\rho}\frac{\overline{u}^2 + \overline{v}^2}{2} + \hat{\rho}c_p T\right) + \nabla_h \cdot \left(\hat{\rho}\mathbf{u}\frac{\overline{u}^2 + \overline{v}^2}{2} + \hat{\rho}\mathbf{u}\frac{2\kappa - 1}{\kappa}c_p\overline{T}\right) = 0. \quad (3)$$

Finally, averaging the equation of state, we find that $\hat{p} = \hat{\rho}R\overline{T}$. The equation for the kinetic energy of vertically averaged motion follows from the first two Equations (2)

$$\frac{\partial}{\partial t}\hat{\rho}\frac{\overline{u}^2 + \overline{v}^2}{2} + \frac{\partial}{\partial x}\hat{\rho}\overline{u}\frac{\overline{u}^2 + \overline{v}^2}{2} + \frac{\partial}{\partial y}\hat{\rho}\overline{v}\frac{\overline{u}^2 + \overline{v}^2}{2} = -\overline{u}\frac{\partial\hat{p}}{\partial x} - \overline{v}\frac{\partial\hat{p}}{\partial y}.$$

When combining the resulting equation with Equation (3), we shall have

$$\frac{D}{Dt}\hat{p} + \frac{2\kappa - 1}{\kappa}\hat{p}\left(\frac{\partial\overline{u}}{\partial x} + \frac{\partial\overline{v}}{\partial y}\right) = 0 ,$$

where the symbol

$$\frac{D}{Dt} = \frac{\partial}{\partial t} + \overline{u}\frac{\partial}{\partial x} + \overline{v}\frac{\partial}{\partial y}$$

for the two-dimensional material time-derivative is introduced. As a result, we arrive at a closed system of equations of motion of a two-dimensional, compressible, baroclinic fluid film in the field of the Coriolis force, which was originally derived by Alishayev (1980):[1]

---

[1] A two-dimensional baroclinic atmosphere model over the topography with height h has been considered in Kurgansky et al. (1996). For that model, under quasi-geostrophic approximation (see Section 2.3) it has been shown that the energy integral coincides with the sum of kinetic and free energy (a specific form of available enthalpy) from Section 1.5.

$$\frac{D}{Dt}\bar{u} = -\frac{1}{\hat{\rho}}\frac{\partial\hat{p}}{\partial x} + l\bar{v}, \quad \frac{D}{Dt}\bar{v} = -\frac{1}{\hat{\rho}}\frac{\partial\hat{p}}{\partial y} - l\bar{u},$$

$$\frac{D\hat{\rho}}{Dt} + \hat{\rho}\left(\frac{\partial\bar{u}}{\partial x} + \frac{\partial\bar{v}}{\partial y}\right) = 0,$$

$$\frac{D}{Dt}\ln\hat{p} - \frac{2\kappa-1}{\kappa}\frac{D}{Dt}\ln\hat{\rho} = 0. \tag{4}$$

Equations (4) differ from the purely two-dimensional equations (when $w \equiv 0$ in the entire atmospheric volume) in the respect that in the thermodynamic equation one finds the quantity $(2\kappa - 1)/\kappa = 9/7$, instead of the heat capacity ratio $\kappa = c_p/c_v = 7/5 > 9/7$, which holds for two-atomic molecules of the gas of which the Earth's atmosphere practically fully consists.[2] Thus, Alishayev's model realizes a lesser degree of horizontal elasticity of a fluid as compared with an ordinary two-atomic gas. Correspondingly, a lower speed of sound is realized at the same absolute temperature, this being due to the fact that Equations (4) implicitly allow the vertical air motion.

Equations (4) may be deduced by a simpler method than that demonstrated above if we recall that fluid dynamic equations could be closed either by the energy equation or entropy equation (Landau and Lifshitz, 1988, paragraph 49). It should be recalled that in meteorology it is more convenient to use the potential temperature θ instead of the entropy. From the adiabaticity condition $D\theta/Dt = 0$, it follows after averaging over height

$$\frac{\partial}{\partial t}\hat{\rho}\bar{\theta} + \frac{\partial}{\partial x}\hat{\rho}\overline{u\theta} + \frac{\partial}{\partial y}\hat{\rho}\overline{v\theta} = 0.$$

---

[2] After averaging over height, the adiabatic index transforms according to the rule $\kappa \to \kappa^* = (2\kappa - 1)/\kappa$. Note that $\kappa^* - \kappa = (\kappa - 1)^2/\kappa > 0$ and $\kappa^* = \kappa$ if and only if $\kappa = 1$, i.e., the process is isothermal. The transformation rule written above has a solution $\kappa = f(m) = (2m + 3)/(2m + 1)$, where m is an integer. It means that when $\kappa = f(m)$, $\kappa^* = f(m + 1)$. For instance, if $\kappa = 5/3$, $\kappa^* = 7/5$ and if $\kappa = 7/5$, $\kappa^* = 9/7$, and so on. We conclude that the three-dimensional quasi-static dynamics of an m-atomic gaseous atmosphere is reduced to the two-dimensional dynamics of an $m + 1$-atomic gaseous atmosphere. Suppose additionally that the gas molecular weight μ and the gas constant R are the constants of the transformation in question: $R = R^*$. In this case, the specific heat capacity at constant pressure transforms according to the rule: $c_p = R\kappa/(\kappa - 1) \to c_p^* = R^*\kappa^*/(\kappa^* - 1) \equiv R(2\kappa - 1)/(\kappa - 1)$. As the result, $c_v^* = c_p^*/\kappa^* = c_p$, the following statement holds: the total potential energy of a three-dimensional quasi-static atmosphere after averaging transforms into the internal energy of a two-dimensional atmosphere.

We assume that the potential temperature does not change with altitude $\theta = \overline{\theta} = \theta(x, y, t)$ and get

$$\frac{\partial}{\partial t} \hat{\rho}\overline{\theta} + \frac{\partial}{\partial x} \hat{\rho}\overline{u}\,\overline{\theta} + \frac{\partial}{\partial y} \hat{\rho}\overline{v}\,\overline{\theta} = 0.$$

Finally, we determine the functional dependence between variables $\theta, \hat{\rho}$ and $\hat{p}$:

$$\hat{p} = \int\limits_0^\infty p\,dz = \int\limits_0^\infty \rho R T\,dz = \int\limits_0^\infty \rho R \theta \left(p/p_{00}\right)^{(\kappa-1)/\kappa} dz$$

$$= \frac{R\theta}{g} \int\limits_0^{p_0} \left(p/p_{00}\right)^{(\kappa-1)/\kappa} dp = \frac{R\theta}{g} p_{00} \frac{\kappa}{2\kappa-1} \left(\hat{\rho}/\hat{\rho}_{00}\right)^{(2\kappa-1)/\kappa} \quad (5)$$

with $\hat{\rho}_{00} = p_{00}/g$. Based on this, we arrive at the last Equation (4).

The formulated atmospheric model contains the effect of horizontal baroclinicity, which has been taken into account in a classical paper by Blinova (1943). What is the horizontal baroclinicity itself? It should be recalled that a baroclinic fluid is a fluid with not coinciding equiscalar surfaces of any two thermodynamic variables: p and ρ, p and T, etc. When intersecting, these isoscalar surfaces generate families of thermodynamic solenoids (tubes) which, according to the well-known Bjerknes' theorem (Haltiner and Martin, 1957; Haltiner and Williams, 1980; Pedlosky, 1987), determine the vorticity genesis in the atmosphere. When a family of horizontal planes cuts the solenoids in question into cells of finite volume, this property is attributed to the horizontal baroclinicity of the atmosphere. In the framework of reduced Equations (4) the effect of horizontal baroclinicity appears as the lack of coincidence between isolines $\hat{\rho} = \text{const}$ and $\hat{p} = \text{const}$. That is why the model is able to reproduce the genesis of vorticity in the atmosphere. For instance, the problem of zonal flow baroclinic instability could be set and effectively solved (Alishayev, 1980).[3] Nevertheless, the physical interpretation of the results obtained

---

[3] A one-level non-divergent model of the atmosphere with horizontal buoyancy which imitates the baroclinic effects has been developed by Tennekes (1977), who gives a physical interpretation of the zonal flow instability mechanism resulting in this model. The mean western zonal wind is maintained by thermal forcing. Relatively warm (and light) air parcels deviate poleward due to the Coriolis force, and relatively cold (and heavy) air parcels deviate equatorward. This is the case of the so-called large-scale convection in the field of the Coriolis force.

provides certain difficulties, mainly when it concerns small-scale perturbations. Theory predicts their unlimited growth – something similar to 'the ultraviolet catastrophe' (in the theory of black-body radiation) is observed. It is necessary to note that the time rate of perturbation growth reaches saturation when the spatial scale decreases, which makes the analogy not perfect. A more correct account of the vertical baroclinicity of the atmosphere and the allowance for the updraft and downset motions, supporting the geostrophic balance, enables one to arrive at a qualitatively correct baroclinic instability diagram of a zonal flow with no amplifying small-scale perturbations (see Chapter 3).[4]

Fundamentally important results are obtained in a special case of a barotropic atmosphere when the latter is horizontally compressible, but the variables $\hat{\rho}$ and $\hat{p}$ are in one-to-one functional dependence. For example, it is sufficient to demand that the atmosphere has a spatially uniform potential temperature within its entire volume. We introduce a barotropic potential as a new variable

$$\hat{\phi} = \int_{\hat{\rho}_{00}}^{\hat{\rho}} \left(\hat{\rho}(\hat{p}) \right)^{-1} d\hat{p} \ .$$

After that, using a chain of equalities

$$d\hat{\phi} = \frac{d\hat{p}}{\hat{\rho}} = \frac{d\hat{p}}{d\hat{\rho}} d \ln \hat{\rho} = c^2 \ln \hat{\rho}$$

we reduce Equations (4) to the system

$$\frac{D}{Dt}\bar{u} = -\frac{\partial \hat{\phi}}{\partial x} + l\bar{v}, \quad \frac{D}{Dt}\bar{v} = -\frac{\partial \hat{\phi}}{\partial y} - l\bar{u},$$

---

[4] Equations (4) have been obtained by averaging the three-dimensional, quasi-static hydrothermodynamic equations over height, assuming that the horizontal wind and potential temperature do not significantly change in the vertical direction. This rough approximation disregards the coupling between the horizontal and vertical components of motion resulting from the thermal wind relation. Strictly speaking, Equations (4) should be considered as a correction to barotropic equations, taking into account small effects of horizontal inhomogeneity in the temperature field. Nevertheless, they give surprisingly good results, as was already mentioned by Tennekes (1977). A simple climate model constructed based on similar principles and successfully applied for the global and regional climate peculiarities modeling is described in Petoukhov et al. (2000).

$$\frac{D\hat{\phi}}{Dt} + c^2 \left(\frac{\partial \bar{u}}{\partial x} + \frac{\partial \bar{v}}{\partial y} \right) = 0. \tag{6}$$

Assuming c^2 to be a constant quantity equal to c_0^2, one arrives at a system of equations written down by Obukhov (1949a)

$$\frac{D}{Dt} \bar{u} = -\frac{\partial \hat{\phi}}{\partial x} + l\bar{v}, \quad \frac{D}{Dt} \bar{v} = -\frac{\partial \hat{\phi}}{\partial y} - l\bar{u},$$

$$\frac{D\hat{\phi}}{Dt} + c_0^2 \left(\frac{\partial \bar{u}}{\partial x} + \frac{\partial \bar{v}}{\partial y} \right) = 0. \tag{7}$$

Here $c_0^2 = RT_0$ is the isothermal (Newtonian) speed of sound estimated for the average surface air temperature value. Under real conditions, one has $T_0 \approx 288$ K and $c_0 \approx 280$ m·s^{-1}. Let us perform the formal limit transition to the case of incompressible fluid $\kappa = c_p/c_v \rightarrow \infty$ in Equations (4). Instead of the last Equation (4) one will have

$$\frac{D}{Dt} \ln \hat{p} - 2 \frac{D}{Dt} \ln \hat{\rho} = 0$$

or

$$\frac{D}{Dt} \ln \left(\hat{p} \hat{\rho}^{-2} \right) = 0.$$

This is the model of incompressible but horizontally heterogeneous fluid, when $\rho = \rho(x,y,t)$. The height of the fluid free surface $h = h(x,y,t)$ is related to the bottom pressure $p_0(x,y,t)$ according to the equation $h = p_0/\rho g$. The resulting system of equations formally coincides with the equations for two-dimensional adiabatic motion of a perfect gas with the polytropic index $n = 2$ (cf. Landau and Lifshitz, 1988, paragraph 53). The most fundamental is the case of a homogeneous fluid. Here, instead of (5), one has

$$\hat{p} = \int_0^h p\,dz = \int_0^{p_0} z\,dp = \int_0^h z\rho g\,dz = g\rho \frac{h^2}{2} = \frac{g}{2\rho} \hat{\rho}^2 \ .$$

The result are the so called 'shallow water' equations

$$\frac{D}{Dt}\bar{u} = -g\frac{\partial h}{\partial x} + l\bar{v}, \quad \frac{D}{Dt}\bar{v} = -g\frac{\partial h}{\partial y} - l\bar{u},$$

$$\frac{D}{Dt}\ln h + \frac{\partial \bar{u}}{\partial x} + \frac{\partial \bar{v}}{\partial y} = 0. \tag{8}$$

The quantity

$$c^2 = \frac{d\hat{p}}{d\hat{\rho}} = d\left(\frac{g\hat{\rho}^2}{2\rho}\right)\bigg/ d\hat{\rho} = \frac{g\hat{\rho}}{\rho} = gh$$

is an analog of the speed of sound in acoustics. In the framework of this model, c has the simple meaning of the velocity of propagation for long gravity waves ('shallow water waves'). This velocity coincides with the speed of sound propagation in the barotropic atmospheric model if the condition $gh = RT_0$ holds, i.e., the mean depth of the fluid layer in the shallow water model is equal to the atmospheric scale height $H_0 = RT_0/g \approx$ 8 km.

Subjecting the first two Equations (7) to cross-differentiation in order to eliminate the potential $\hat{\phi}$, one has

$$\frac{\partial}{\partial x}\left(\frac{D\bar{v}}{Dt}\right) - \frac{\partial}{\partial y}\left(\frac{D\bar{u}}{Dt}\right) = -l\left(\frac{\partial \bar{u}}{\partial x} + \frac{\partial \bar{v}}{\partial y}\right) - \bar{v}\frac{dl}{dy},$$

where it is assumed that the Coriolis parameter l depends on the latitudinal coordinate y. Writing down the left-hand side of this equation and cancelling the similar terms, we obtain

$$\frac{D}{Dt}\left(\frac{\partial \bar{v}}{\partial x} - \frac{\partial \bar{u}}{\partial y}\right) + \left(\frac{\partial \bar{v}}{\partial x} - \frac{\partial \bar{u}}{\partial y}\right)\left(\frac{\partial \bar{u}}{\partial x} + \frac{\partial \bar{v}}{\partial y}\right) = -l\left(\frac{\partial \bar{u}}{\partial x} + \frac{\partial \bar{v}}{\partial y}\right) - \frac{Dl}{Dt}.$$

Eliminating the two-dimensional divergence $(\partial\bar{u}/\partial x) + (\partial\bar{v}/\partial y)$ from the above written equation and the continuity equation, we arrive at the equation of potential vorticity conservation

$$\frac{D}{Dt}\frac{(\partial\bar{v}/\partial x) - (\partial\bar{u}/\partial y) + l}{\hat{\rho}} = 0 \tag{9}$$

discovered by Rossby (1940) for the shallow water model (8), and later obtained by Obukhov (1949a) for the barotropic atmospheric model (7). It looks as if the term 'potential vorticity' has been proposed by Rossby and Obukhov independently of each other, most probably, by analogy with the potential temperature. The 'potentiality' for the creation of vorticity due to changes in the mass $\hat{\rho}$ of a fluid column of a unit horizontal area is clearly expressed by formula (9). Equation (9) is the fluid dynamic interpretation of the principle of angular momentum conservation applied to a cylinder with vertical walls and the base area σ, cut off from the fluid. In the course of fluid motion the mass of this cylinder $m = \hat{\rho}\sigma$ is, first, time-constant. Second, the angular momentum proportional to the product of vorticity (doubled angular velocity of a fluid local rotation) on the inertia momentum of the cylinder is conserved. It means that the ratio of the angular momentum to the mass squared is invariant. This is just the potential vorticity conservation law.

In the framework of a simple model (7) the meaning of the vorticity charge concept introduced in Chapter 1 becomes evident. If in a Cartesian plane a material domain Σ is given, which is bounded by a material closed curve, the vorticity charge of a fluid enclosed within Σ is expressed by the integral

$$Z = \iint_{\Sigma} \left(\frac{\partial \bar{v}}{\partial x} - \frac{\partial \bar{u}}{\partial y} + l \right) dxdy = \text{invariant}$$

which coincides with the absolute vorticity flux across Σ. Invariant Z is independent of the invariant of the mass of fluid

$$M = \iint_{\Sigma} \hat{\rho} dxdy \, .$$

It is natural to attribute the potential vorticity value averaged over Σ to the ratio Z/M.

An analogy exists between the dynamics of Ertel's potential vorticity on isentropic surfaces in a three-dimensional baroclinic atmosphere and two-dimensional barotropic fluid dynamics, the latter being realized in the case of a rapid background rotation. In the two-dimensional model we are free from the necessity to account for the influence of baroclinic factors which, though permanently present in the atmosphere, often play an indirect role. This role mainly reduces to the generation of vortices within narrow spatial and temporal intervals. In the framework of a barotropic model this could be replaced by the action of some effective external source of vorticity. On the other hand, when solving the modern problems, withstanding the theory of atmospheric motion predictability, the nonlinear dynamic processes adequately described by equations similar to Equations (7, 8) happen to be the most important.

A close analogy between the two-dimensional barotropic atmospheric dynamics (7) and shallow water dynamics in the field of the Coriolis force (8) had been made the basis of effective methods of modeling of large-scale atmospheric processes using rotating laboratory setups (Lorenz, 1967; Hide and Mason, 1975; Dolzhanskii and Golitsyn, 1977). The most delicate point is how to fulfil the requirements of the criteria of similarity when performing such a modelling, particularly with account for viscous dissipation.

2.2   Adjustment of fluid dynamic fields

Let us first consider a barotropic horizontally incompressible atmosphere moving on an l-plane, i.e., when $l = \text{const}$. With the account for the incompressibility of motion, the stream function $\overline{\psi}$ is introduced such that $\overline{u} = -\partial \overline{\psi}/\partial y$, $\overline{v} = \partial \overline{\psi}/\partial x$ and the Coriolis force in Equations (7) of Section 2.1 becomes a potential force, having the potential $l\,\overline{\psi}$. As it follows from the vorticity equation, the motion of such an incompressible uniformly rotating fluid is dynamically indistinguishable from that of a fluid without background rotation. The only difference is that the potential $\hat{\phi}$ for a rotating fluid is related to that $\hat{\phi}_0$ for a non-rotating fluid as

$$\hat{\phi} = \hat{\phi}_0 + l\,\overline{\psi}, \tag{1}$$

i.e., in the rotating frame of reference a strong linear dependence between the pressure field $\hat{\phi}$ and the velocity (stream function) field $\overline{\psi}$ appears. We shall recall that the potential $\hat{\phi}_0$ satisfies Poisson's equation

$$\nabla^2 \hat{\phi}_0 = 2J\left(\frac{\partial \overline{\psi}}{\partial x}, \frac{\partial \overline{\psi}}{\partial y}\right),$$

where J denotes the Jacobian operator. When the rotation is rapid, one arrives at the equation of approximate geostrophic balance

$$\hat{\phi} \approx l\,\overline{\psi}, \tag{2}$$

which holds to the accuracy of the terms dropped out in (1). In fact, for large-scale atmospheric processes, this is the accuracy of 10%. The geostrophic wind equations

$$\bar{u} = -l^{-1} \partial\hat{\phi}/\partial y, \ \bar{v} = l^{-1} \partial\hat{\phi}/\partial x$$

are the direct consequence of Equation (2).

The geostrophic wind is directed along the isobars. In the Northern Hemisphere, if one stands with one's back to the wind, the pressure increases in the right-hand side direction in order to compensate for the deflective Coriolis force (Buys Ballot law).[5] For real air flows the geostrophic wind equation is satisfied approximately. In this case, air motion is called quasi-geostrophic.

Several general questions arise. Why the wind is close to the geostrophic wind, nevertheless? How, taking into account the atmospheric horizontal compressibility, one can explain the mutual adjustment between the pressure and velocity fields? What will happen further on, if the wind deviates from its geostrophic value at the initial time? It is difficult to solve a non-linear, non-stationary problem, which arises when one attempts to answer the above questions. It is possible to consider a small part of the atmosphere with a slight imbalance between the wind and pressure fields and to use the procedure of linearization of governing equations.

The problem of the adjustment of oceanic (marine) fields had been formulated by Rossby (1938), and in a one-dimensional case had been studied by Cahn (1945). The most complete solution for the barotropic atmosphere belongs to Obukhov (1949a). Account for the atmospheric vertical stratification, i.e., for the entropy variations with height, does not bring conceptually new difficulties into the adjustment problem. An additional class of internal gravity waves appears, and the adjustment of meteorological fields in the main bulk of the atmosphere occurs through the emission of these waves. This takes place only after the vertically averaged meteorological characteristics due to purely barotropic mechanisms have been adjusted, and somehow lengthens the adjustment process.

As a background atmospheric state we consider the uniform geostrophic air motion with straight line isobars, along which air parcels steadily move without acceleration $(D\bar{u}/Dt = D\bar{v}/Dt = 0)$. Averaging symbols in the equations of the previous section will be omitted further, and the background atmospheric state is marked by subscript 'g'. This state is superimposed by small perturbations $u = u_g + u'$, $v = v_g + v'$, $\phi = \phi_g + \phi'$, and $lu_g = -\partial\phi/\partial y$, $lv_g = \partial\phi/\partial x$, the Coriolis parameter being considered as a constant. Without the loss of generality, because of the Galilean principle, it can be

---

[5] Professor Buys Ballot of Utrecht wrote in 1857: 'If in the Northern hemisphere you stand with your back to the wind, pressure is lower on your left hand than on your right. In the Southern hemisphere the reverse is true' (see Wallace and Hobbs, 1977).

assumed that $u_g = v_g = 0$. Linear (with respect to the perturbations) equations (7) of Section 2.1 take the form

$$\frac{\partial u'}{\partial t} = -\frac{\partial \phi'}{\partial x} + lv', \quad \frac{\partial v'}{\partial t} = -\frac{\partial \phi'}{\partial y} - lu',$$

$$\frac{\partial \phi'}{\partial t} + c_0^2 \left(\frac{\partial u'}{\partial x} + \frac{\partial v'}{\partial y} \right) = 0. \tag{3}$$

We take divergence and curl operations from the first two Equations (3), respectively:

$$\frac{\partial}{\partial t} \left(\frac{\partial u'}{\partial x} + \frac{\partial v'}{\partial y} \right) = -\nabla^2 \phi' + l \left(\frac{\partial v'}{\partial x} - \frac{\partial u'}{\partial y} \right),$$

$$\frac{\partial}{\partial t} \left(\frac{\partial v'}{\partial x} - \frac{\partial u'}{\partial y} \right) = -l \left(\frac{\partial u'}{\partial x} + \frac{\partial v'}{\partial y} \right).$$

When combining the second resulting equation with the third equation (3), we arrive at the local conservation law

$$\frac{\partial}{\partial t} \tilde{\Omega} = 0, \quad \tilde{\Omega} = \frac{\partial v'}{\partial x} - \frac{\partial u'}{\partial y} - \frac{l}{c_0^2} \phi', \tag{4}$$

which stands for the linear version of the theorem on potential vorticity conservation (see Equation (9) in Section 2.1). The conservation law (4) has been obtained for linear equations only, but this is actually a much stronger statement if compared with the potential vorticity conservation law in a general nonlinear case. In a linear problem, there is eternal local memory, and the initial values $\tilde{\Omega}(x, y)$ are remembered forever. The existence of the invariant $\tilde{\Omega}(x, y)$ gives one a possibility to classify the solutions of Equations (3) in a general way. There are, first, stationary solutions of Equations (3) with non-vanishing field $\tilde{\Omega}$ and, second, the wave solutions, for which $\tilde{\Omega} \equiv 0$ by definition.

With the account for the compressibility of motion we decompose the velocity field onto the sum of solenoidal and potential components

$$u' = -\frac{\partial \psi'}{\partial y} + \frac{\partial \phi'}{\partial x}, \quad v' = \frac{\partial \psi'}{\partial x} + \frac{\partial \phi'}{\partial y},$$

where ψ' is the stream function and φ' is the velocity potential. The divergence and vorticity equations are written as

$$\frac{\partial}{\partial t} \nabla^2 \varphi' = -\nabla^2 \phi' + l\nabla^2 \psi', \quad \frac{\partial}{\partial t} \nabla^2 \psi' = -l\nabla^2 \varphi'.$$

A benefit of the transition $(u',v') \rightarrow (\psi',\varphi')$ is due to the fact that it is possible to reduce the order of equations by 'dropping away' the Laplace operator ∇^2 from both their sides, because the solutions, being harmonic functions, are assumed to be regular at infinity. As a result, we arrive at the equations

$$\frac{\partial \varphi}{\partial t} = -\phi + l\psi, \quad \frac{\partial \psi}{\partial t} = -l\varphi,$$

$$\frac{\partial \phi}{\partial t} + c_0^2 \nabla^2 \varphi = 0, \tag{5}$$

where the primes over the variables are omitted. Equations (5) contain two classes of solutions. First, the stationary solutions which describe the geostrophic balance between the velocity and pressure fields $l\psi_s = \phi_s$, $\varphi_s = 0$ (cf. Equation (2)). System (5) is degenerative for these solutions. From the third equation (5) in a stationary case it follows that $\nabla^2 \varphi_s = 0$, which is a simple consequence of the second equation. At the same time, one of the variables, either ψ_s or ϕ_s, stays arbitrary. These stationary solutions are characterized by non-zero values of the invariant $\tilde{\Omega}$

$$\tilde{\Omega}_s = \nabla^2 \psi_s - \frac{l}{c_0^2}\phi_s \equiv \nabla^2 \psi_s - \frac{1}{L_0^2}\psi_s,$$

where $L_0 = c_0/l$ is some characteristic length-scale. Second, there are non-stationary or wave solutions of Equations (5) which are gauged by the condition of the invariant (4) vanishing

$$\tilde{\Omega}_w = \nabla^2 \psi_w - \frac{l}{c_0^2}\phi_w \equiv 0. \tag{6}$$

When such calibration is used, each variable $\zeta_w = (\varphi_w,\psi_w,\phi_w)$ satisfies the wave equation of Klein–Gordon type

$$\frac{\partial^2}{\partial t^2}\zeta_w + l^2\zeta_w = c_0^2\nabla^2\zeta_w, \tag{7}$$

which has a particular solution in the form of a harmonic wave

$$\zeta_w = A\exp\{i(\mathbf{k}\cdot\mathbf{x} - \sigma t)\},$$

where $\mathbf{k}$ is the two-dimensional wave vector and σ is frequency, if and only if the characteristic equation $\sigma^2 = l^2 + \mathbf{k}^2 c_0^2$ holds. Thus, the waves have the lowest limit frequency l, which coincides with that of inertial oscillations.

One could arrive at Equation (6) more easily. We start from Equation (4) and search for its wave solutions, with harmonic time-dependence $\exp\{-i\sigma t\}$, the frequency σ being non-zero. Substituting this solution into Equation (4), we get $-i\sigma\tilde{\Omega}_w = 0$. As $\sigma \neq 0$, one has with necessity that $\tilde{\Omega}_w = 0$.

Let us formulate the initial, or Cauchy, problem for Equations (5). Assume that at the initial time, $t = 0$, the hydrodynamic fields are geostrophically balanced everywhere on an unbounded plane except of some domain of diameter $2R$ much smaller than the length scale L_0. In this way, non-zero initial conditions

$$\varphi = -l^{-1}\frac{\partial\psi(x, y, 0)}{\partial t} = f(x, y, 0),$$

$$\frac{\partial\varphi}{\partial t} = -\phi(x, y, 0) + l\psi(x, y, 0) = g(x, y, 0)$$

are specified inside a circle of radius R for the wave equation (cf. Equation (7))

$$\frac{\partial^2}{\partial t^2}\varphi + l^2\varphi = c_0^2\nabla^2\varphi$$

that results from Equations (5). We shall not write down the explicit form of the solution of the resulting problem, referring the reader to the corresponding mathematical literature (e.g., Courant and Hilbert, 1953) as well as to the original paper by Obukhov (1949a). Instead, some qualitative arguments, based on the energy conservation principle, will be proposed, which lead to the same ultimate result. The reader eager to study the adjustment theory at first hand, will find a reproduction of the paper by Obukhov (1949a) in the book (Obukhov, 1988). A detailed account of the results gained in the adjustment theory for the baroclinic atmosphere could be found in the textbook by Monin (1990).

The energy conservation law for Equations (5) reads as

$$\frac{\partial}{\partial t}\left\{\frac{u^2+v^2}{2}+\frac{1}{2c_0^2}\phi^2\right\}+\frac{\partial}{\partial x}(u\phi)+\frac{\partial}{\partial y}(v\phi)=0$$

which, along with the regularity of solutions at infinity, results in the total energy conservation

$$E=\iint\left(\frac{u^2+v^2}{2}+\frac{1}{2c_0^2}\phi^2\right)dxdy$$

with an integration extended over the entire unlimited plane. We decompose the hydrodynamic fields onto the sum of stationary and wave components: $u=u_s+u_w$, $v=v_s+v_w$, $\phi=\phi_s+\phi_w$, with $lu_s=-\partial\phi_s/\partial y$, $lv_s=\partial\phi_s/\partial x$. In such a way, the energy is split onto three terms

$$E=\iint\left\{\frac{1}{2}\left(\nabla\psi_s\right)^2+\frac{1}{2L_0^2}\psi_s^2\right\}dxdy$$

$$+\iint\left\{\frac{\partial\psi_s}{\partial x}v_w-\frac{\partial\psi_s}{\partial y}u_w+\frac{1}{c_0^2}l\psi_s\phi_w\right\}dxdy$$

$$+\iint\left\{\frac{1}{2}\left(u_w^2+v_w^2\right)+\frac{1}{2c_0^2}\phi_w^2\right\}dxdy.$$

The second integral in this sum is transformed by integration by parts

$$\iint\left\{\frac{\partial}{\partial x}(\psi_s v_w)-\frac{\partial}{\partial y}(\psi_s u_w)-\left(\frac{\partial v_w}{\partial x}-\frac{\partial u_w}{\partial y}-\frac{l}{c_0^2}\phi_w\right)\psi_s\right\}dxdy\ .$$

Under imposed boundary conditions at infinity, it vanishes for an arbitrary function ψ_s if and only if $\tilde{\Omega}_w\equiv0$. Now, the energy is equal to the sum of energies of the stationary and wave components: $E=E_s+E_w$. In our problem the energy of the wave component at the initial moment of time $t = 0$ is concentrated within a circle of radius R. If A_0 is the amplitude of the function $f(x,y,0)$, then the estimate holds

$$E_w\big|_{t=0} \propto \pi R^2 A_0^2 .$$

As time goes on, E_w dissipates within a circle of gradually increasing radius $R(t) \propto c_0 t$. Based on the energy conservation law (with the account for E_s = invariant), we shall have

$$E_w \propto \pi c_0^2 t^2 A^2(t) = \text{invariant},$$

which immediately leads to the conclusion that the amplitude of velocity potential $A(t)$ decreases in time according to the law $A(t) \propto (1/t)$.

The adjustment process is characterized by the time-scale $\tau = 2R/c_0$. If we take $R = 500$ km and $c_0 = 280$ m$\cdot$s^{-1}, then $\tau \approx 1$ h. A more thorough analysis shows that after a time of 3–4 h the geostrophic balance between the fields in question is practically re-established inside the domain with the initial filed imbalance. The wind field is uniquely determined through the theorem on the conservation of $\tilde{\Omega}$

$$\nabla^2 \psi_s - \frac{1}{L_0^2} \psi_s = \tilde{\Omega}(x, y, 0). \tag{8}$$

One could arrive at condition (8) starting from a variational principle proposed by Diky (1969). It is postulated that the adjustment process occurs in such a way that the outgoing wave has the least energy among all possible differences between the initial and the final stationary state. The minimum of the functional

$$E_w = \iint \frac{1}{2}\left\{\left(u + \frac{\partial \psi_s}{\partial y}\right)^2 + \left(v - \frac{\partial \psi_s}{\partial x}\right)^2 + \frac{1}{c_0^2}(\phi - l\psi_s)^2\right\} dxdy$$

is searched for by giving variations to the stream function ψ_s of a stationary state, towards which the adjustment occurs. The first variation δE_w has the form'

$$\delta E_w = \iint \left\{\left(u + \frac{\partial \psi_s}{\partial y}\right)\frac{\partial \delta \psi_s}{\partial y} - \left(v - \frac{\partial \psi_s}{\partial x}\right)\frac{\partial \delta \psi_s}{\partial x}\right.$$

$$-\frac{1}{c_0^2}(\phi - l\psi_s)l\delta\psi_s\bigg\}dxdy\,.$$

Integration by parts gives

$$\delta E_w = \iint\left\{\left(\frac{\partial v}{\partial x} - \frac{\partial u}{\partial y} - \frac{l}{c_0^2}\phi - \nabla^2\psi_s + \frac{1}{L_0^2}\psi_s\right)\delta\psi_s\right\}dxdy\,.$$

Because the variation $\delta\psi_s$ is arbitrary, Equation (8) follows immediately from the requirement that $\delta E_w = 0$. It is easily checked that E_w is minimal when Equation (8) holds.

The equation

$$\nabla^2\psi_s - \frac{1}{L_0^2}\psi_s = \kappa\delta(r),$$

where $\delta(r)$ is a two-dimensional delta function, has the solution

$$\psi_s = -\frac{\kappa}{2\pi}K_0\left(\frac{r}{L_0}\right) \tag{9}$$

which serves as the fundamental solution of Equation (8). With its help, and based on a superposition principle, a general solution of Equation (8) can be constructed, for an arbitrary function in its right-hand side

$$\psi_s(\mathbf{x}) = -\frac{1}{2\pi}\iint K_0\left(\frac{\rho}{L_0}\right)\tilde{\Omega}(\mathbf{x}',0)d\mathbf{x}',$$
$$\rho^2 = (\mathbf{x} - \mathbf{x}')^2.$$

Using a well-known asymptotics for the McDonald function $K_0(r/L_0)$ valid at $r \gg L_0$, we shall have

$$\psi_s \approx -\frac{\kappa}{2}\frac{1}{\sqrt{2\pi(r/L_0)}}\exp\left\{-\frac{r}{L_0}\right\}.$$

Thus, scale L_0 has the meaning of a radius of screening of a point singularity in the field of invariant $\tilde{\Omega}$. In the Russian-language literature, L_0 is called the Obukhov's synoptic scale, or Rossby–Obukhov's radius of deformation. In mid latitudes, $L_0 \approx 3000$ km. When $r << L_0$, or formally in the case of a two-dimensionally incompressible atmosphere, with $L_0 \to \infty$, the Green's function (9) transforms into $\psi = (\kappa/2\pi)\ln r$, which refers to a concentrated (point) Helmholtz' vortex in two-dimensional incompressible fluid dynamics (Batchelor, 1967), being the fundamental solution of the Poisson's equation for a stream function over an unbounded domain.

The concept of a singular geostrophic vortex (see Equation (9)) has been laid into the basis of investigation on the dynamics of a system of point vortices described by Kirchhoff's equations, with the Hamiltonian function containing $-K_0(r/L_0)$ instead of $\ln r$ (see, e.g., Gryanik, 1983a). When the baroclinicity effects are taken into account, the Rossby's internal radius of deformation $L_1 = NH/l$ becomes essential. Here $H = c_0^2/g$ is the scale height and $N = 10^{-2}\ s^{-1}$ is the Brunt–Vasala frequency, which is the principal characteristic of the atmospheric vertical stratification. For the Earth's atmosphere, $(L_1/L_0)^2 \approx 0.1$. It is of interest to note that the singular geostrophic vortices have been modelled in a laboratory (Griffiths and Hopfinger, 1986).

The most important external similarity criterion for planetary atmospheres is the ratio of a planet radius a to the scale L_0, which enables one to classify the planets of the Solar System according to the types of general circulation of their atmospheres (Golitsyn, 1973). Earth and Mars, with $a/L_0 \geq 1$, are moderate rapidly rotating planets. Here the radius of the domain of influence of a point source in the potential vorticity field is limited from above by the scale L_0. Jupiter and Saturn, with $a/L_0 >> 1$, are rapidly rotating planets. Venus, with $a/L_0 << 1$, is a slowly rotating planet, when the domain of influence of a point source in the potential vorticity field is limited from above by the planetary size.

In the framework of a laboratory experiment, modelling the geophysical fluid flows, the external geometrical scale of a laboratory setup, R, often plays the role of the planet radius a. For instance, it could be the width of a channel if the experiment is carried out in an annulus. In particular, when $R/L_0 \leq 1$, the flow patterns similar to the motions observed in the Earth's atmosphere are reproduced (see, e.g., Dolzhanskii et al., 1979). In these experiments the external scale R determines the characteristic spatial scale of motion. When $R/L_0 >> 1$, one can expect that the laboratory experiment would reproduce the characteristic features of atmospheric circulation on giant planets (see Nezlin, 1986).

Adjustment of meteorological fields: an example. We have shown that energy E could be split into the sum of energy of the adjusted state E_s and wave energy E_w. Besides, energy is divided into kinetic and potential energy. If at the initial time the pressure field is spatially uniform, and the velocity field is determined by the stream function ψ_0, the initial potential energy is zero. If the radius of the initial vortex is small as compared with Obukhov's synoptic scale L_0, the potential energy P equal to the difference between the kinetic energies of the initial and final states is approximately equally divided between the adjusted state and the wave component. To prove it, we compose the difference between the kinetic energies of the initial and ultimate states, denoting the function of the latter stream as ψ_s:

$$P = \iint \frac{1}{2}\left(\nabla\psi_0\right)^2 dxdy - \iint \frac{1}{2}\left(\nabla\psi_s\right)^2 dxdy.$$

Using the Green's formula and the regularity of fields at infinity, we shall have

$$P = -\iint \frac{1}{2}\left\{\left(\psi_0 + \psi_s\right)\left(\nabla^2\psi_0 - \nabla^2\psi_s\right)\right\}dxdy.$$

The stream function ψ_s is determined by the equation

$$\nabla^2\psi_s - \frac{1}{L_0^2}\psi_s = \nabla^2\psi_0.$$

For motions with a spatial scale much smaller than L_0 one has $\psi_s \approx \psi_0$ and

$$P \approx -\iint \psi_s\left(\nabla^2\psi_0 - \nabla^2\psi_s\right)dxdy = 2\iint \frac{1}{2L_0^2}\psi_s^2 dxdy,$$

which proves the above statement. In this case the velocity field has changed insignificantly, but the pressure field has been transformed dramatically. Now, instead of the vanishing values $\phi_0 = 0$, one has $\phi_s = l\psi_s/c_0^2 \cong l\psi_0/c_0^2$. Thus, the pressure field has been adjusted to the velocity field. As the kinetic energy is one order of magnitude larger than the potential energy, it is the pressure field that transforms during the adjustment process. In this way, the smallest energy is spent for the generation of the outgoing wave.

Consider a simple example when

$$\psi_0(x,y) = A_0 J_0(r/R), \quad r^2 = x^2 + y^2.$$

Here J_0 is the Bessel function of the zeroth order, $R \ll L_0$ is the vortex spatial scale and A_0/R is the wind speed scale. The stream function ψ_s is calculated simply as

$$\psi_s(x,y) = A_s J_0(r/R), \quad A_s = A_0 L_0^2/\left(L_0^2 + R^2\right).$$

Therefore

$$P = \left\{ 1 - \frac{L_0^4}{\left(L_0^2 + R^2\right)^2} \right\} \iint \frac{1}{2} \left(\nabla \psi_0\right)^2 dxdy$$

$$\approx 2 \frac{R^2}{L_0^2} \iint \frac{1}{2} \left(\nabla \psi_0\right)^2 dxdy.$$

On the other hand, the potential (elastic) energy of the adjusted state is estimated by the integral

$$P_s = \iint \frac{1}{2L_0^2} \psi_s^2 dxdy = \frac{L_0^4}{\left(L_0^2 + R^2\right)^2} \iint \frac{1}{2L_0^2} \psi_0^2 dxdy$$

$$= \frac{L_0^2 R^2}{\left(L_0^2 + R^2\right)^2} \iint \frac{1}{2} \left(\nabla \psi_0\right)^2 dxdy,$$

so that $P \approx 2P_s$.

2.3   Filtered (quasi-geostrophic) equations

In the previous section, we considered in detail the problem of the adjustment of hydrodynamic fields for a barotropic atmosphere model. In the framework of a linear problem the exact geostrophic balance re-establishes at the expense of emission of waves transporting excess energy to infinity. This process takes a time not exceeding the value $\tau \sim l^{-1} = 10^4$ s = 3 h. Non-linear processes, as well as the effect of Earth's sphericity (the latter appearing in the form of latitudinal Coriolis parameter changes, though weak, for spatial scales considered), being ignored in such a treatment, permanently tend to turn the atmosphere away from the state of exact geostrophic balance. However, the time necessary for these mechanisms to work efficiently, is one order of magnitude greater than the characteristic time-scale of the adjustment process. From this point of view the latter is a rapid process. As the result, the atmosphere never deviates noticeably from the adjusted state, and the ratios of the geostrophic wind 'work well' to the accuracy of about 10%.

Thus, from the standpoint of physics there are two characteristic time-scales inherent in our problem: (i) the fast time-scale τ, which is the characteristic time-scale of the adjustment process, and (ii) the slow time-scale T due to the perturbative influence of both non-linearities and beta-effect (as meteorologists call the effect of the variability of the Coriolis parameter with latitude in fluid dynamic equations). Our further goal is to arrive at self-consistent equations for slow evolution, with the characteristic time-scale T, of adjusted atmospheric states. Equations we are searching for should describe only the meteorologically significant motions and not to contain (filter out) the 'wave noise'.

First, it is necessary to reduce Equations (7) of Section 2.1 to a non-dimensional form convenient for analysis. We would write these equations once more, omitting all the symbols above the variables

$$\frac{Du}{Dt} - lv = -\frac{\partial \phi'}{\partial x}, \quad \frac{Dv}{Dt} + lu = -\frac{\partial \phi'}{\partial y},$$

$$\frac{D\phi'}{Dt} + c_0^2 \left(\frac{\partial u}{\partial x} + \frac{\partial v}{\partial y} \right) = 0. \tag{1}$$

We introduce the wind speed scale U, spatial scale L, slow time-scale $T = L/U$ and determine the dimensionless variables

$$(\bar{u}, \bar{v}) = \frac{1}{U}(u, v), \quad (\bar{x}, \bar{y}) = \frac{1}{L}(x, y),$$

$$\bar{t} = \frac{U}{L} t.$$

The Coriolis parameter l varies with latitude y and has a characteristic (average) value $\tilde{l}$, which is used to determine the dimensionless Coriolis parameter $\bar{l} = l/\tilde{l}$. The ϕ'-scale is estimated on the basis of the geostrophic balance relation so that $\bar{\phi'} = \phi'/\tilde{l} UL$. When reduced to a dimensionless form, Equations (1) become

$$\varepsilon \left(\frac{\partial \bar{u}}{\partial t} + \bar{u} \frac{\partial \bar{u}}{\partial x} + \bar{v} \frac{\partial \bar{u}}{\partial y} \right) = -\frac{\partial \bar{\phi'}}{\partial x} + \bar{l} \bar{v} \quad,$$

$$\varepsilon \left(\frac{\partial \bar{v}}{\partial t} + \bar{u} \frac{\partial \bar{v}}{\partial x} + \bar{v} \frac{\partial \bar{v}}{\partial y} \right) = -\frac{\partial \bar{\phi'}}{\partial y} - \bar{l} \bar{u} \quad,$$

$$\beta^2 \varepsilon \left(\frac{\partial \bar{\phi'}}{\partial t} + \bar{u} \frac{\partial \bar{\phi'}}{\partial x} + \bar{v} \frac{\partial \bar{\phi'}}{\partial y} \right) + \frac{\partial \bar{u}}{\partial x} + \frac{\partial \bar{v}}{\partial y} = 0. \tag{2}$$

Here, because $U = 10$ m·s^{-1} and $L = 10^6$ m for large-scale processes, the small Kibel'–Rossby parameter $\varepsilon = U/\tilde{l}L = 0.1$ and the small parameter of large-scale atmospheric compressibility $\beta^2 = (L/L_0)^2 = (L\tilde{l}/c_0)^2 = 0.1$ appear. For motions with the spatial scale L comparable to the Obukhov's synoptic scale L_0 the parameters ε and $\varepsilon\beta^2$ are of the same order of magnitude. It is assumed that $\tilde{l} = 1 + \varepsilon\lambda$, where the variable λ is of the order of unity, i.e., the Coriolis parameter λ varies by the value of εl over the spatial scale L.

We seek the solution of Equations (2) in the form of expansion onto powers of the small parameter ε: $u = u_0 + \varepsilon u_1 + \dots$, $v = v_0 + \varepsilon v_1 + \dots$, $\phi' = \phi'_0 + \varepsilon\phi'_1 + \dots$. The bar above the variables is omitted hereafter. In the zeroth order of ε one has

$$-\partial\phi'_0/\partial x + v_0 = 0, \quad -\partial\phi'_0/\partial y - u_0 = 0, \quad \partial u_0/\partial x + \partial v_0/\partial y = 0, \qquad (3)$$

i.e., the velocity field of the zeroth order of approximation is solenoidal and has a stream function, which coincides with ϕ'_0. Let us mention that the system (3) is degenerative: the third equation is the direct consequence of the first two equations and, simultaneously, one variable remains arbitrary. It is convenient to assume that this is the function ϕ'_0. Further on, ϕ'_0 is defined by the condition of resolvability of equations for the first order of approximation on ε. This is the idea underlying the quasi-geostrophic expansion. Starting from the resulting first approximation equations it is easy to arrive at the vorticity equation

$$\frac{D^0}{Dt}\nabla^2\phi'_0 = -\left(\frac{\partial u_1}{\partial x} + \frac{\partial v_1}{\partial y}\right) - \frac{d\lambda}{dy}\frac{\partial\phi'_0}{\partial x},$$

where

$$\frac{D^0}{Dt} = \frac{\partial}{\partial t} + u_0\frac{\partial}{\partial x} + v_0\frac{\partial}{\partial y} \equiv \frac{\partial}{\partial t} - \frac{\partial\phi'_0}{\partial y}\frac{\partial}{\partial x} + \frac{\partial\phi'_0}{\partial x}\frac{\partial}{\partial y}.$$

By eliminating the two-dimensional velocity divergence from this equation as well as from the continuity equation taken at the same first approximation, we get the closed evolutionary equation for the field $\phi'_0(x,y,t)$:

$$\frac{D^0}{Dt}\left(\nabla^2\phi'_0 - \beta^2\phi'_0 + \lambda\right) = 0,$$

having the form of a material conservation law. Returning back to the dimensional variables and introducing the geostrophic stream function $\psi =$

$\phi'/\tilde{l}$, we finally arrive at the theorem on potential vorticity conservation in quasi-geostrophic approximation

$$\frac{\partial}{\partial t}\left(\nabla^2\psi - \frac{1}{L_0^2}\psi\right) + J\left(\psi, \nabla^2\psi + l\right) = 0, \qquad (4)$$

where J denotes the Jacobian operator

$$J(A, B) = \frac{\partial A}{\partial x}\frac{\partial B}{\partial y} - \frac{\partial A}{\partial y}\frac{\partial B}{\partial x}.$$

Equation (4) has been derived by A. M. Obukhov (1949) in the neglect of the beta-effect though a possibility of its account has been mentioned. Virtually at the same time, J. Charney (1948), using a scaling analysis, established an equation very close to Equation (4). In notations used in this Section, Charney's (1948) equation reads as

$$\frac{D}{Dt}(\zeta + l) = \frac{1}{c_0^2}(\zeta + l)\frac{\partial\phi'}{\partial t}, \quad \zeta = \frac{1}{l}\nabla^2\phi'.$$

That is why Equation (4) is often referred to as the Charney–Obukhov equation.[6] In fact, this is one of the fundamental equations of modern dynamic meteorology and geophysical fluid dynamics, which is definitely of everlasting theoretical significance.

It is the easiest way to derive the corresponding equation for the baroclinic atmosphere, with the account for altitudinal field dependence, by using isobaric coordinates, with pressure p becoming an independent variable instead of altitude. Under quasi-static approximation, the use of such coordinates ($\partial p/\partial z = -\rho g$) greatly simplifies the form of equations of motion and continuity. The procedure of the transition to p-coordinates is thoroughly described in (Kibel', 1957; Thompson, 1961; Ryym, 1990). The idea is that for every scalar function f the identity $f(x,y,z,t) = f\{x,y,p(x,y,t),t\}$ could be written down. By differentiating both of its parts with respect to the

---

[6] Ertel (1941) was the first to derive the equation $\partial\nabla^2\psi/\partial t + J(\psi, \nabla^2\psi + l) = 0$, which differs from Equation (4) by the absence of the term $-L_0^2(\partial\psi/\partial t)$, which accounts for the atmospheric large-scale compressibility. When deriving this equation, Ertel (1941) deliberately neglected the horizontal wind divergence term lD, because he assumed that for synoptic-scale motions magn $(lD) = (10^{-1}–10^{-2})$ magn $(\partial\nabla^2\psi/\partial t)$, though nowadays one might not agree with the lower boundary of this estimate.

arguments x,y,z, and t, the relations we seek are easily derived. Below, we give the equations of horizontal motion and continuity:

$$\frac{Du}{Dt} + w*\frac{\partial u}{\partial p} - lv = -g\frac{\partial z}{\partial x},$$

$$\frac{Dv}{Dt} + w*\frac{\partial v}{\partial p} + lu = -g\frac{\partial z}{\partial y},$$

$$\frac{\partial u}{\partial x} + \frac{\partial v}{\partial y} + \frac{\partial w*}{\partial p} = 0$$

without derivation. Here, the geopotential height $z(x,y,p,t)$ of an isobaric surface becomes a dependent function. We should recall that gz is named the geopotential and is equal to the amount of work which should be performed in order to replace a unit air mass from the Earth's surface to a given baric level. Besides, $w* = dp/dt$ (which is an analog of vertical velocity) also becomes a dependent variable, and

$$\frac{D}{Dt} = \frac{\partial}{\partial t} + u\frac{\partial}{\partial x} + v\frac{\partial}{\partial y}$$

is the symbol of material time-derivative taken along an isobaric surface. All the derivatives with respect to horizontal coordinates and time are taken at p = const but, as before, u and v remain the genuine horizontal wind components. The thermodynamic equation is taken in an approximate form, analogous to that used for simplification of the continuity equation in the shallow water model

$$gp^2 \frac{D}{Dt}\frac{\partial z}{\partial p} + \alpha^2 c_0^2 w* = 0,$$

where $c_0^2 = RT_0$ again. Here T_0 is the average surface air temperature, and the notation α^2 is adopted for the baroclinicity parameter $\dfrac{R(\gamma_a - \gamma)}{g}\dfrac{T}{T_0}$ ($\gamma_a = g/c_p \approx 10$ K km^{-1} is the dry adiabatic lapse rate) which, generally speaking, is allowed to be a weak function of altitude (pressure) at the expense of corresponding variations of values of temperature lapse rate γ and temperature T averaged over isobaric surfaces. For simplicity, it is assumed that $\alpha^2 \approx 0.1$ is a quasi-constant.

Asymptotic expansion of the above written equations with respect to the powers of the small Kibel'–Rossby parameter, practically analogous to that made for the barotropic atmosphere, enables one to arrive at the conservation law for quasi-geostrophic potential vorticity:

$$\frac{\partial}{\partial t}q + J(\psi,q) = 0,$$

$$q = \nabla^2\psi + l + \frac{\partial}{\partial p}m^2 p^2 \frac{\partial\psi}{\partial p}, \qquad (5)$$

where the parameter $m^2 = L_0^{-2}\alpha^{-2}$ has the dimension of a wave number squared. The quantity $L_1 = m^{-1}$ is the internal Rossby radius of deformation and its value is close to 10^6 m. The scale L_1 has the meaning of the radius of influence of a point singularity in the potential vorticity field (a point 'vorticity charge'). In the case of $l = $ const, the dynamics of the point vorticity charges is governed by equations given in (Gryanik, 1983b), where the fluid motion stream function is expressed through the Green's function of a boundary problem which could be posed for Equation (5). Equation (5) is written in terms of a geostrophic stream function $\psi = gz/\tilde{l}$. It is of the second order with respect to the variable p and should be supplied by the proper boundary conditions at the upper and lower boundaries of the atmosphere. Customarily, it is assumed that

$$\frac{\alpha^2 c_0^2 w *}{g} = -\frac{D}{Dt}p^2\frac{\partial\psi}{\partial p} \to 0,$$

at $p \to 0$. There are more difficulties associated with the lower boundary. Strictly speaking, the boundary condition has to be written at the surface $z(x,y,p,t) = 0$, specified by the solution itself, which extremely complicates the problem. That is why, in practice, this condition is approximately taken at the isobaric level $p = p_0$, which in the best way, e.g., in the least squares sense, fits the Earth's surface. Now, from the thermodynamic equation and the vertical velocity definition in p-coordinates

$$w = \frac{D}{Dt}z + w *\frac{\partial z}{\partial p},$$

under the condition of the Earth's surface impermeability, and using the hydrostatic equation in order to estimate $\partial z/\partial p$ at $p = p_0$, the following boundary condition follows

$$\frac{\partial}{\partial t}\Theta + J(\psi,\Theta) = 0, \quad \Theta = -p\frac{\partial\psi}{\partial p} - \alpha^2\psi, \quad p = p_0, \tag{6}$$

where the variable Θ is proportional to the deviation of the surface air potential temperature from its value averaged over the Earth's surface.

2.4   Rossby waves

Dynamic meteorology also uses more accurate approximations than the quasi-geostrophic approximation. One of the appropriate methods to improve the accuracy of quasi-geostrophic equations is to use a quasi-solenoidal approximation, which results from the expansion of dependent variables into asymptotic sets arranged with respect to the powers of the Mach number squared. The principal part of the velocity field becomes its solenoidal component $u = -\partial\psi/\partial y$, $v = \partial\psi/\partial x$. The potential vorticity conservation law for the atmospheric barotropic model, described by Equations (7) of Section 2.1, is reduced to the relation

$$\frac{\partial}{\partial t}q + J(\psi,q) = 0, \quad q = \ln(\nabla^2\psi + l) - \phi/c_0^2, \tag{1}$$

and the linkage between the fields ψ and ϕ is given by the balance equation

$$\nabla^2\phi = \nabla(l\nabla\psi) + 2J\left(\frac{\partial\psi}{\partial x}, \frac{\partial\psi}{\partial y}\right),$$

which appears, by the way, as the second approximation equation in quasi-geostrophic expansion of Equations (7) in Section 2.1. The relation between ψ and ϕ, equivalent to the linear balance equation, has been established by Blinova (1943).

Equations of quasi-solenoidal approximation are valid both near the equator, where the Coriolis parameter l is small and quasi-geostrophic approximation fails, and in the extratropics, where the Kibel'–Rossby number ε is small. In extratropics, Equation (1) holds with accuracy $O(\varepsilon^2)$, the accuracy of the quasi-geostrophic equation being only $O(\varepsilon)$ in this case. However, the reconstruction of the velocity field starting from the known potential vorticity q-distribution meets here with the well-known mathematical difficulties (Bolin, 1955; Charney, 1955; Monin, 1972).

Mention should also be made of semi-geostrophic equations (Hoskins, 1975; Blumen, 1981), where an increase in accuracy by one order of magnitude is achieved as compared to the quasi-geostrophic case (under the limitation of l = const), at the expense of coordinate transformation according to the rule

$$(x, y) \rightarrow (x + v_g/l, y - u_g/l),$$

where u_g and v_g are geostrophic wind components.[7] This transformation has been suggested by Yudin (1955).

The case of an incompressible atmosphere ($c_0^2 = \infty$) is conceptually much simpler. Here Equation (1) reduces to the conservation law for the absolute vorticity vertical component, and it is not necessary any more to consider the balance equation. In spherical coordinates, with the poles situated on the Earth's rotation axis, this equation takes the form

$$\frac{\partial}{\partial t} \nabla^2 \psi + J(\psi, \nabla^2 \psi + 2\Omega \cos \vartheta) = 0, \tag{2}$$

where ∇^2 is the Laplace operator on the sphere (sometimes, called the Legendre operator, which is the angular-dependent part of the true Laplace operator), and the brackets denote the Jacobian operator

$$J(A, B) = \frac{1}{a^2 \sin \vartheta} \left(\frac{\partial A}{\partial \vartheta} \frac{\partial B}{\partial \lambda} - \frac{\partial A}{\partial \lambda} \frac{\partial B}{\partial \vartheta} \right).$$

Here λ is the longitude, ϑ is the co-latitude (equal to $\pi/2 - \phi$, where ϕ is the latitude), and a is the Earth's radius.

In the absence of rotation, when $\Omega = 0$, Equation (2) is invariant with respect to a group of arbitrary rotations of the sphere. Solution of Equation (2) can be represented by an expansion onto spherical harmonics

$$Y_n^m(\vartheta, \lambda) = e^{im\lambda} P_n^m(\cos \vartheta),$$

---

[7] Later on, using Hamiltonian methods, semi-geostrophic equations have been derived for a general case, free of limitation l = const (see Salmon, 1998).

where $P_n{}^m$ are associated Legendre polynomials of the first kind. A set of functions $Y_n{}^m$ with the fixed lower index n (degree) forms an elementary state (subspace) of dimension $2n + 1$, which is invariant with respect to a group of arbitrary solid-body-like rotations of the sphere. Each point of this subspace corresponds to a Laplace operator eigenfunction, with the eigenvalue $-n(n + 1)/a^2$. The simplest situation, with $n = 1$, corresponds to an ordinary three-dimensional space with basis functions $Y_1{}^0$, $Y_1{}^1$ and $Y_1{}^{-1}$. Every function from this space accounts for a fluid solid-body-like rotation on the sphere about an axis, which is arbitrarily directed respective to the spherical coordinate axes. If this axis is directed along the axis of a fluid rotation, then the fluid motion pattern is described by the spherical harmonic $Y_1^0 = P_1(\cos \vartheta) = \cos \vartheta$.

If we take an arbitrary Laplace operator eigenfunction, which obligatorily belongs to a certain elementary state, then it will be an exact stationary solution of the Helmholtz equation (2), when $\Omega = 0$, because of the well-known anti-symmetric property of Jacobian operators: $J(A,B) = -J(B,A)$. In the presence of background rotation, these stationary solutions are 'animated', and the ψ_n-function, which belongs to an elementary state with index n (with $Y_n{}^m$ as the basis functions), gains an explicit time dependence

$$\psi_n = \psi_n\left(\vartheta, \lambda - \omega_n t\right),$$

where

$$\omega_n = -\frac{2\Omega}{n(n+1)}.$$

In other words, the entire pattern of isolines $\psi_n = $ const starts to rotate, in solid-body-like manner, with the angular velocity ω_n, directed opposite to the angular velocity Ω of the background rotation and proportional by magnitude to Ω (vanishing simultaneously with Ω), and monotonically decreasing along with index n increasing. We have obtained a solution of atmospheric dynamic equations which corresponds to nonlinear Rossby–Haurwitz–Blinova waves (Rossby et al., 1939; Haurwitz, 1940; Blinova, 1943).

When the atmosphere rotates in zonal direction like a solid body with an angular velocity $\alpha\Omega$, Equation (2) has the exact solution

$$\tilde{\psi}_n\left(\vartheta, \lambda - \omega_n t\right) = -\alpha\Omega a^2 Y_1^0 + \psi_n\left(\vartheta, \lambda - \omega_n t\right),$$

where

$$\omega_n = \alpha\Omega - \frac{2\Omega(1+\alpha)}{n(n+1)}.$$

Isolines $\tilde{\psi}$ = const become stationary relative to the Earth's surface provided $\alpha = 2/[n(n+1) - 2]$. That is why one can observe semi-permanent centers of action in the atmosphere, which are the vast baric minima and maxima occupying a stationary position over the Earth's surface in the presence of westerlies (Rossby et al., 1939).

In the most general case, when the vector of a fluid solid-body rotation angular velocity is not directed along the vector $\vec{\Omega}$, an exact time-dependent solution of the Helmholtz equation on a sphere is pointed out in Rochas (1986). As special cases, it contains all the exact solutions of Equation (2), derived earlier in Ertel (1943) (see also references in Kochin et al., 1964); Craig, 1945; Blinova, 1946; Neamtan, 1946; Thompson, 1982; Verkley, 1984.

If the internal (Newtonian) viscosity is taken into account, one has to introduce the term $\nu\nabla^2(\nabla^2 + 2/a^2)\psi$ into the right-hand side of Equation (2). Here ν is the coefficient of kinematic fluid viscosity. The mathematical form of this viscous term is uniquely dictated, along with the account of the differential operator order, by: (i) the requirement of the viscous operator covariance with respect to the group of solid-body-like rotations of a sphere (any polynomial of ∇^2 satisfies this requirement), and (ii) by the requirement that viscosity should not influence the fluid solid-body-like rotation, i.e., the viscous operator must vanish for every eigenfunction from the elementary state $n = 1$.

Account for the atmospheric large-scale compressibility ($c_0{}^2 \neq \infty$) results in a dramatic complication of the theory. A structure of small-amplitude oscillations about the atmospheric state of rest is described by eigenfunctions of the Laplace's tidal operator, which are called the Hough functions (Hough, 1898). In the general case these functions are expressed in the form of infinite superposition of spherical harmonics belonging to different elementary states. One can find a summary of Hough function theory and their applications in Longuet-Higgins (1968), Diky (1969), and also Lindzen (1990). We only notice that this theory essentially includes the parameter $\gamma = 4\Omega^2 a^2/c_0{}^2$, which makes Hough functions not universal, contrary to the spherical harmonics.[8] Linear Rossby waves are the asymptotic (at $\gamma \to 0$) solutions of the Laplace's tidal equation, under the simultaneous requirement that the ratio of wave frequency to Ω remains finite.

---

[8] Negative $\gamma = 4\Omega^2 a^2/gh$ values (h is a dynamically equivalent fluid depth) must be included in a complete calculation of the response of the atmosphere to external forces (in the tidal theory).

2.5   Continuous (spontaneous) wave emission

We should distinguish between spontaneous emission of rapid waves, accompanying continuous adjustment of meteorological fields towards a more balanced atmospheric state, and emission as the result of unbalanced initial conditions. The latter is the subject of the Rossby's adjustment problem considered in Section 2.2. In any case, the adjustment process is accompanied with energy changes as well as with air mass local redistribution. McIntyre and Norton (1989)[9] have proposed a general scheme of finding the balanced states, when the potential vorticity field is known, which is based on the principle of a minimum of spontaneous wave emission. It is worth mentioning that the extrapolation of the well-known Lighthill's aerodynamic theory of sound wave generation (see, e.g., Crighton, 1981) to the case of non-zero fluid background rotation needs additional studies. In connection with this problem, an idealized example of spontaneous wave emission by a pair of singular vortices within a rotating fluid is given at the end of this Section.[10]

A nonlinear system of governing equations for the shallow water model on a unbounded l-plane is considered:

$$\frac{Du}{Dt} - lv = -\frac{\partial \phi}{\partial x}, \quad \frac{Dv}{Dt} + lu = -\frac{\partial \phi}{\partial y},$$

$$\frac{D}{Dt} \ln \phi + \frac{\partial u}{\partial x} + \frac{\partial v}{\partial y} = 0, \quad \phi = gh.$$

These equations are re-written in equivalent terms

---

[9] The full publication of McIntyre and Norton (1989) results in the open literature had been delayed until 2000 (see McIntyre and Norton, 2000). Below, all the corresponding references are given to McIntyre and Norton (2000).

[10] The current status of the problem is characterized in Ford *et al.* (2000) (see also McIntyre (2001). From this paper a reader will learn about conceptual difficulties associated with the concept of 'slow invariant manifold'. This is an invariant functional subspace, on which the atmospheric dynamics is governed by potential vorticity conservation principle solely, and the information on potential vorticity field alone completely characterizes the atmospheric state, without inertia-gravity waves. The paper cited contains an interesting discussion of the issue of the impossibility for a *strict slow manifold* to exist, because of a loss of time-reversibility property owing to inevitably present irreversible spontaneous wave emission. From this viewpoint, our example of wave emission by two point vortices, with an asymptotic limit state of no wave radiation also deserves certain attention.

$$\frac{\partial u}{\partial t} - lv + \frac{\partial \phi}{\partial x} = -N_u, \quad \frac{\partial v}{\partial t} + lu + \frac{\partial \phi}{\partial y} = -N_v,$$

$$\frac{\partial \phi}{\partial t} + \phi_0 \left(\frac{\partial u}{\partial x} + \frac{\partial v}{\partial y} \right) = -N_\phi,$$

where the notations

$$\phi_0 = c_0^2, \quad N_u = (\mathbf{v} \cdot \nabla)u, \quad N_v = (\mathbf{v} \cdot \nabla)v,$$

$$N_\phi = \mathbf{v} \cdot \nabla \phi + (\phi - \phi_0)(\nabla \cdot \mathbf{v}) = \nabla \cdot \left((\phi - \phi_0)\mathbf{v} \right)$$

are used. Finally, introducing new dependent variables of vorticity $\zeta = (\partial v/\partial x) - (\partial u/\partial y)$ and velocity divergence $\delta = (\partial u/\partial x) + (\partial v/\partial y)$, we consider the problem

$$\frac{\partial \zeta}{\partial t} + l\delta = -N_\zeta, \tag{1}$$

$$\frac{\partial \delta}{\partial t} - l\zeta + \nabla^2 \phi' = -N_\delta, \tag{2}$$

$$\frac{\partial \phi'}{\partial t} + c_0^2 \delta = -N_\phi. \tag{3}$$

Here, $\phi' = \phi - \phi_0$, $N_\zeta = \nabla \cdot (\mathbf{v}\zeta)$, $N_\delta = \nabla \cdot ((\mathbf{v} \cdot \nabla)\mathbf{v})$, $N_\phi = \nabla \cdot (\mathbf{v}\phi')$. A heterogeneous wave equation for the divergence δ follows from Equations (1)–(3)

$$\left(\frac{\partial^2}{\partial t^2} + l^2 - c_0^2 \nabla^2 \right) \delta = -\nabla \cdot \left\{ l\zeta \mathbf{v} + \frac{\partial}{\partial t}((\mathbf{v} \cdot \nabla)\mathbf{v}) - \nabla^2 (\mathbf{v}\phi') \right\}. \tag{4}$$

We split the velocity field into the sum of solenoidal and divergent components: $\mathbf{v} = \mathbf{v}_\psi + \mathbf{v}_\varphi$, $\nabla \cdot \mathbf{v}_\psi = 0$, $\nabla \times \mathbf{v}_\varphi = 0$. The underlying idea is that the resulting system would account for a slow temporal evolution of dependent variables at the expense of time changes in $\tilde{\Omega} = \zeta - l\phi'/c_0^2$, occurring according to the equation

$$\partial \tilde{\Omega}/\partial t + \nabla \cdot (\mathbf{v}\tilde{\Omega}) = 0.$$

The quantity $\tilde{\Omega}$ is related to the deviation of the potential vorticity $I = (\zeta + l)/(c_0^2 + \phi')$ from its reference value $I_0 = l/c_0^2$ described by the exact formula $I - I_0 = \tilde{\Omega}/(c_0^2 + \phi')$. All the other fields must be expressed diagnostically through $\tilde{\Omega}$.

The simplest way to derive the desired approximate equations, which is the 'first-order inversion' by terminology adopted in McIntyre and Norton (2000), has been proposed by Charney (1955) and consists in the full neglect of the divergent velocity component as compared with the solenoidal component. In other words, both divergence δ and its first derivative with respect to time $\partial\delta/\partial t$ are taken to equal zero. Instead of Equations (2), we now have a nonlinear balance equation

$$-l\zeta + \nabla^2\phi' = -\nabla \cdot \left\{ \left(\mathbf{v}_\psi \cdot \nabla \right) \mathbf{v}_\psi \right\}.$$

This equation has to be solved together with the equation

$$\zeta - \frac{l}{c_0^2}\phi' = \left(I - I_0 \right)\left(c_0^2 + \phi' \right),$$

where $I(\mathbf{x},t)$ is a given function.

'Second-order direct inversion', proposed by McIntyre and Norton (2000), consists in the use of heterogeneous wave Equation (4), in the neglect of the divergence second time derivative $\partial^2\delta/\partial t^2$ in its left-hand side:

$$\left(l^2 - c_0^2\nabla^2 \right)\delta = -\nabla \cdot \left\{ l\zeta\mathbf{v} + \frac{\partial}{\partial t}((\mathbf{v} \cdot \nabla)\mathbf{v}) - \nabla^2\left(\mathbf{v}\phi' \right) \right\}. \qquad (5)$$

Further on, this equation will be used instead of Equation (3). Moreover, like in the 'first-order inversion', the first time derivative $\partial\delta/\partial t$ in Equation (2) is taken to equal zero:

$$-l\zeta + \nabla^2\phi' = -\nabla \cdot \left\{ (\mathbf{v} \cdot \nabla)\mathbf{v} \right\}.$$

Lynch (1989) has derived his 'slow equations' by neglecting the term $\partial\{(\mathbf{v}\cdot\nabla)\mathbf{v}\}/\partial t$ in the right-hand side of Equation (5). McIntyre and Norton (2000) have chosen another route. The local time derivative of velocity, $\partial\mathbf{v}/\partial t$, is replaced by an auxiliary diagnostical variable $\mathbf{a}$, with the dimension of acceleration, which is closely related to the ideas suggested earlier in the framework of the initialization theory. Now, Equation (5) takes the form

$$\left(l^2 - c_0^2\nabla^2\right)\delta = -\nabla\cdot\left\{l\zeta\mathbf{v} + (\mathbf{a}\cdot\nabla)\mathbf{v} + (\mathbf{v}\cdot\nabla)\mathbf{a} - \nabla^2\left(\mathbf{v}\phi'\right)\right\}.$$

In order to arrive at the closed system of equations, one needs to use Equation (1) once more, also substituting $\mathbf{a}$ for $\partial\mathbf{v}/\partial t$ there:

$$\nabla\times\mathbf{a} + l\delta = -N_\zeta.$$

The proposed procedure can be generalized towards a 'third-order inversion', etc. when the derivatives $\partial^2\delta/\partial t^2$, $\partial^3\delta/\partial t^3$, etc. are taken to equal zero but, at the same time, the non-vanishing values of the time derivatives of a lower degree are taken into account. The method of deriving these equations is rather simple. To arrive at a more accurate approximation one needs additional time differentiations which increase the formal order of equations. The emerging derivatives $\partial\mathbf{v}/\partial t$ are replaced in a regular manner with auxiliary diagnostic variables $\mathbf{a}_1$, $\mathbf{a}_2$, etc. The scheme proposed in McIntyre and Norton (2000) does not exactly conserve energy and momentum. Neither it allows the local conservation of mass, because all these quantities vary in the course of adjustment. Only the total atmospheric mass is constant. This procedure had been tested for the shallow water equations over a hemisphere (the problem was solved by the iteration method, using the inversion schemes of both second and third order) and had shown a remarkably good agreement with the results of direct integration of primitive Equations (8) of Section 2.1. This indicates that the information accumulated in the potential vorticity field is virtually exhaustive for meteorological purposes.

From the standpoint of a refined fluid dynamic theory, two promising directions for the development of the initialization procedure could be drawn. The first direction is related to the use of the Hough function, when initial fields are projected onto rotational Hough modes only. These modes correspond to the Rossby waves in the asymptotic case of incompressible atmosphere (see Section 2.4). Following the second direction, the potential vorticity concept is applied, and initial fields are concorded with the help of a routine of potential vorticity field computation and subsequent inversion. Recently, in the course of a practical realization of the initialization procedure, an integration of governing equations with time-increments directed forward and backward in time has been commonly used. Because of the irreversibility of both diabatic heating and friction, one should introduce a special correction to compensate for the systematic action of these factors. A corresponding procedure is called the diabatic initialization.

Example of spontaneous wave emission. As a wave source, consider two singular point vortices, with intensities $\Gamma_1 > 0$ and $\Gamma_2 > 0$, which are separated at a distance $2a$ from each other and permanently rotate about the center of vorticity with the angular velocity $\omega = (\Gamma_1 + \Gamma_2)/8\pi a^2$. The validity of the system of inequalities is assumed

$$4\pi a^2 l \leq \Gamma_1 + \Gamma_2 \ll 8\pi a c_0, \tag{6}$$

which means that, on the one hand, the angular velocity of the rotation of the vortices exceeds the vertical component of the Earth's rotation $l/2$, but, on the other hand, Mach number $M = (\Gamma_1 + \Gamma_2)/8\pi a c_0$ is small, i.e., the approximation of a nearly incompressible fluid is used.

Axial coordinates (r, ϑ) are introduced, with the origin of co-ordinates being placed in the center of vorticity that lies on a straight line connecting the vortices. At distances $r \sim a$ the atmosphere can be treated as horizontally incompressible. Outside the vortices, the vorticity field is potential, the time-variable part of a velocity potential φ, induced by the vortices, being given by the formula (see Milne-Thompson, 1960; Batchelor, 1967)

$$\varphi = -\frac{i}{\pi} \frac{\Gamma_1 \Gamma_2}{\Gamma_1 + \Gamma_2} \left(\frac{a}{r}\right)^2 \exp\{2i(\omega t - \vartheta)\}, \tag{7}$$

valid at $a \ll r \ll \pi c_0 \omega^{-1}$.

At large distances from the vortices one has to account for fluid compressibility and, therefore, for its background rotation. However, the velocity field could be estimated under linear approximation. In a wave zone, at $r \geq \pi c_0 \omega^{-1}$, the velocity potential satisfies: (i) the wave equation (7) from Section 2.2; (ii) the wave radiation condition at infinity, and (iii) the condition of matching with (7) at $r \ll \pi c_0 \omega^{-1}$, and is given by the formula

$$\varphi = -\frac{1}{4} \frac{\Gamma_1 \Gamma_2}{\Gamma_1 + \Gamma_2} a^2 k^2 H_2^{(2)}(kr) \exp\{2i(\omega t - \vartheta)\}.$$

Here $H_2^{(2)}(kr)$ is the Hankel function of the second kind and $k^2 = (4\omega^2 - l^2)/c_0^2$. A flux of energy, transported by waves at infinity across the circumference of a circle of a sufficiently large radius R, is given by the integral

$$I = \frac{1}{4} \int_0^{2\pi} \left(\frac{\partial \varphi^*}{\partial r} \phi + \frac{\partial \varphi}{\partial r} \phi^*\right) r d\vartheta,$$

where the asterisk denotes the complex conjugate. Using an asymptotic of the Hankel function at $kr \gg 1$ and formula (5) of Section 2.2 which gives the relation between ϕ and φ in the waves, one gets

$$I = \frac{1}{16} \left(\frac{\Gamma_1 \Gamma_2}{\Gamma_1 + \Gamma_2}\right)^2 c_0^2 a^4 \frac{k^6}{\omega}. \tag{8}$$

Without background rotation, at $l = 0$, this formula has been derived in (Klyatskin, 1966; Gryanik, 1983).

Power needed for wave emission is taken from the energy of interaction of the vortices, $E = -(\Gamma_1\Gamma_2/2\pi)\ln 2a$, where a non-essential additive constant is omitted. When losing energy, the vortices diverge and the frequency $\omega = (\Gamma_1 + \Gamma_2)/8\pi a^2$ of emitted waves monotonously decreases. The maximum distance at which the vortices are able to go away from each other is determined through the condition $\omega = l/2$ and is given by the formula $2\tilde{a} = [(\Gamma_1 + \Gamma_2)/\pi l]^{1/2}$. When going away at the distance $2\tilde{a}$ from each other, the vortices do not emit waves. Here the interaction energy is minimum and is equal to $\tilde{E} = -(\Gamma_1\Gamma_2/2\pi)\ln 2\tilde{a}$. Thus, the energy difference

$$\Delta E = E - \tilde{E} = (\Gamma_1\Gamma_2/2\pi)\ln(\tilde{a}/a)$$

characterizes the portion of the vortex interaction energy which is available for the continuous wave emission in an inviscid fluid.

After introducing a new dependent variable $\xi = (a/\tilde{a})^2$, the energy conservation law $d(\Delta E)/dt = -I$ may be written in the form of a differential equation

$$\frac{\xi^2 d\xi}{(1-\xi^2)^3} = -\frac{1}{32\pi}\Gamma_1\Gamma_2\frac{l^3}{c_0^4}dt. \qquad (9)$$

Solution of Equation (9), satisfying the initial condition $\xi(t_0) = \xi_0$, is given by the formula

$$\left[\frac{4\xi}{(1-\xi^2)^2} - \frac{2\xi}{1-\xi^2} - \ln\left|\frac{1+\xi}{1-\xi}\right|\right]_{\xi_0}^{\xi} = \frac{t-t_0}{\tau}, \qquad (10)$$

where $\tau = 2\pi c_0^4/\Gamma_1\Gamma_2 l^3$ appears to be the characteristic time-scale of the vortex system transformation due to wave emission. The time interval τ is extremely large as compared with l^{-1}. In the simplest case of $\Gamma_1 = \Gamma_2 = \Gamma$, $\tau = (L_0/\tilde{a})^4/2\pi l$. Here, L_0 is the Obukhov's synoptic scale. In particular, $\tau \approx 12$ days for $\Gamma = 4\pi\alpha^2 l$, with $\tilde{a}^2 = 2a^2$ and $L_0 = 10\tilde{a}$. According to Equation (10), the square of the distance between the vortices increases initially (when $\xi \ll 1$) following the law (cf. Klyatskin, 1966)

$$\frac{16}{3}(\xi^3 - \xi_0^3) \approx \frac{t-t_0}{\tau}, \qquad (11)$$

and at the final stage, when $\xi \to 1 - 0$, according to the asymptotic formula

$$(1-\xi)^{-2} \approx t/\tau. \qquad (12)$$

From the function $\xi(t)$ found, it is not difficult to calculate the temporal behavior of power emitted by the vortices

$$I = \frac{1}{64\pi} \frac{\Gamma_1 \Gamma_2}{\tau} \frac{\left(1-\xi^2\right)^3}{\xi^3}.$$

According to Equation (11), at $\xi_0 \ll \xi \ll 1$, one has $I \approx (1/12\pi)\Gamma_1\Gamma_2 t^{-1}$, and when $\xi \to 1$, $I \approx (1/8\pi)(\Gamma_1\Gamma_2/\tau)(\tau/t)^{3/2}$, in resemblance to Equation (12).

CHAPTER 3

Hydrodynamic Instability of Conservative Motions

This chapter addresses the problem of large-scale atmospheric motion instability. As a bedrock, the method of non-linear stability analysis with the help of adiabatic motion constants is used. This method proposed by Arnol'd (1965) appears to be a generalization of the well-known Lyapounov's method in the theory of basic stability for the case of fluid dynamics. Problems of atmospheric motion instability are elucidated in more detail in special monographs (Diky, 1976; Drazin and Reid, 1982; Dymnikov and Filatov, 1990; Shakina, 1990) and also in Gill (1982), Pedlosky (1987), Lindzen (1990), Monin (1990), Bluestein (1992), Holton (1992), Cushman-Roisin (1994), and other textbooks.

3.1 Barotropic instability

To clarify the essence of the forthcoming arguments, consider, first, a simple example of the force-free motion of a solid body with a unique fixed point, which coincides with the position of the center of mass. Such a motion is described by the Euler's equations (see, e.g., Landau and Lifshitz, 1973)

$$I_1 \frac{d\omega_1}{dt} = \left(I_3 - I_2\right)\omega_2\omega_3,$$

$$I_2 \frac{d\omega_2}{dt} = \left(I_1 - I_3\right)\omega_3\omega_1,$$

$$I_3 \frac{d\omega_3}{dt} = \left(I_2 - I_1\right)\omega_1\omega_2. \tag{1}$$

Here I_i (i = 1, 2, 3) are the momenta of inertia relative to the principal axes x_i of the solid body; ω_i are the angular velocities of rotation about x_i. For definiteness, it is assumed that $I_1 > I_2 > I_3$.

Equations (1) possess two independent constants of motion: (a) of energy

$$\frac{d}{dt} E = 0, \quad E = I_1 \frac{\omega_1^2}{2} + I_2 \frac{\omega_2^2}{2} + I_3 \frac{\omega_3^2}{2},$$

and (b) of angular momentum squared

$$\frac{d}{dt} M^2 = 0, \quad M^2 = I_1^2 \omega_1^2 + I_2^2 \omega_2^2 + I_3^2 \omega_3^2.$$

Note that Equations (1) describe an inviscid homogeneous fluid motion inside an ellipsoidal cavern, which is called the hydrodynamic gyroscope (Gledzer *et al.*, 1981; Obukhov, 1988). In this case, the constant M^2 has the meaning of a sum of squares of fluid velocity circulation about principal ellipsoid cross-sections. We should also point out that the maximum simplified dynamic equations for the barotropic atmosphere, in the neglect of the beta-effect, could be reduced to Equations (1) (Lorenz, 1960). In this case, M^2 stands for the averaged value of the vorticity squared (enstrophy).

A steady (permanent) rotation about any principal axis of inertia, having an arbitrary angular velocity magnitude, is an equilibrium state of the system described by Equations (1). These permanent rotations appear to be simple analog of the atmospheric zonal circulation. It is convenient to introduce the angular momentum components $M_i = I_i \omega_i$, then the first integrals of Equations (1) are re-written in the form:

$$\frac{M_1^2}{I_1} + \frac{M_2^2}{I_2} + \frac{M_3^2}{I_3} = 2E, \tag{2}$$

$$M_1^2 + M_2^2 + M_3^2 = M^2. \tag{3}$$

In M_i-coordinates, Equations (2, 3) describe the surface of an ellipsoid with semi-axes $(2EI_i)^{1/2}$ and a spherical surface of radius M, respectively. In the course of motion, the edge of the **M**-vector slides along the line of intersection of these surfaces. Evidence of this intersection is provided by the inequalities $2EI_3 < M^2 < 2EI_1$. Geometrically, this means that the radius of the sphere is of an intermediate value with respect to the minor and major ellipsoid semi-axes. In two particular cases, when $M^2 = 2EI_1, 2EI_3$, the sphere and the ellipsoid become tangent in the points lying on the minor or major axis, respectively. These cases correspond to the dynamically stable permanent rotations of the gyroscope about the axis, corresponding to the maximum and minimum momenta of inertia, respectively. In the case of small deviations from these stationary states, the edge of the **M**-vector will move along certain closed curves, which in the infinitesimal limit are the ellipses, encompassing the axes M_1 and M_3 in the neighbourhood of corresponding poles of the ellipsoid. This explains why these permanent rotations are stable. In the case of $M^2 = 2EI_2$, the ellipsoid and the sphere intersect along two circumferences of a big circle crossing each other in the poles of the ellipsoid lying on the M_2-axis. This relates to the unstable permanent rotation about the middle axis. In the case of small perturbations, the trajectories of the **M**-vector edge become hyperbolae and pass away onto large distances from these points.

Thus, qualitatively, using visual geometric arguments (cf. Landau and Lifshitz, 1973; Gledzer *et al.*, 1981) it has been proven that for a permanent gyroscope rotation to be stable, it is necessary and sufficient that energy has an extreme value on the surface, $M^2 = $ const. Let us prove this statement more strictly, so much so that the arguments given below will prove convenient in further discussions of more sophisticated problems of hydrodynamic stability.

Let us assume that the extremum of energy E is sought on condition that $M^2 = $ const. The Lagrange's method of undefined multipliers is used, and a linear combination of the constants of motion $I = E + \lambda M^2$ is composed. Here, λ is a undefined constant multiplier and we seek the absolute extremum for the following expression

$$I = \frac{1}{2}\left(\frac{M_1^2}{I_1} + \frac{M_2^2}{I_2} + \frac{M_3^2}{I_3}\right) + \lambda\left(M_1^2 + M_2^2 + M_3^2\right)$$

in the assumption, which does not restrict the generality, that this extremum is achieved for a permanent rotation about the x_1-axis. Now, one has $\mathbf{M} = \overline{\mathbf{M}} = (\overline{M}_1, 0, 0)$. For a perturbed motion, we have $M_1 = \overline{M}_1 + \delta M_1$, $M_2 = \delta M_2$, $M_3 = \delta M_3$. We decompose I onto a series, which is a finite sum for this particular case, arranged according to the powers of small perturbations of the δM_i magnitude: $I = \overline{I} + \delta I + (1/2)\delta^2 I$, where $\overline{I} = (1/2)\overline{M}_1^2/I_1 + \lambda \overline{M}_1^2$ and $\delta I = (\overline{M}_1/I_1)\delta M_1 + 2\lambda \overline{M}_1\delta M_1$. In order for I to have an extremum when $\mathbf{M} = \overline{\mathbf{M}}$ it is necessary that the first variation δI vanishes: $\overline{M}_1(I_1^{-1} + 2\lambda)\delta M_1 = 0$. Because of this, $\lambda = -1/(2I_1)$ and $\overline{I} = 0$, simultaneously. We write the second variation, using the λ-value found above:

$$\delta^2 I = \left(\frac{1}{I_2} - \frac{1}{I_1} \right)(\delta M_2)^2 + \left(\frac{1}{I_3} - \frac{1}{I_1} \right)(\delta M_3)^2.$$

In the cases of a permanent rotation about the axis with the minimum or maximum momentum of inertia the second variation has a definite sign, i.e., the extremum is actually achieved. Let us prove that these permanent rotations are stable. Suppose for definiteness that the second variation $\delta^2 I$ is positively defined and $I_1 > I_2 > I_3$. We have $I = (1/2)\delta^2 I$, and $\delta^2 I$ is also a constant of motion. If we take $\|\delta \mathbf{M}\|^2 = (\delta M_2)^2 + (\delta M_3)^2$ as a measure of deviations from a stationary state, we immediately have

$$c_1 \|\delta \mathbf{M}\|^2 \le \delta^2 I \le c_2 \|\delta \mathbf{M}\|^2,$$

where $c_1 = I_3^{-1} - I_1^{-1}$ and $c_2 = I_2^{-1} - I_1^{-1}$. Let us choose an arbitrary constant quantity $\varepsilon > 0$. Assume that $\|\delta \mathbf{M}\|^2 \le \delta(\varepsilon)$ at an initial time-instant $t = t_0$. Consequently, $\delta^2 I \le c_2 \delta(\varepsilon)$. However, $\delta^2 I$ is a constant of motion, so $\delta^2 I \le c_2 \delta(\varepsilon)$, and $\|\delta \mathbf{M}\|^2 \le (c_2/c_1)\delta(\varepsilon)$, for any finite time-instant $t = t_0 + \tau$, $\tau > 0$. It is sufficient to choose $\delta(\varepsilon) = (c_1/2c_2)\varepsilon$ in order always to have $\|\delta \mathbf{M}\|^2 < \varepsilon$, which means stability in the Lyapounov's sense (Chetaev, 1990; Dymnikov and Filatov, 1990). We have proved the stability of a permanent rotation about the shortest axis, with the largest momentum of inertia. To establish the stability of a permanent rotation about the longest axis, one should simply consider $-\delta^2 I$, or $|\delta^2 I|$, and perform the same steps as above.

Instability of a permanent rotation about the middle axis is proved with the help of the function $V = \delta M_2 \delta M_3$. For definiteness, it is assumed that $I_3 < I_1 < I_2$. Variations of δM_i obey the equations (cf. Equation (1))

$$I_2 I_3 \frac{d}{dt} \delta M_1 = (I_3 - I_2)\delta M_2 \delta M_3,$$

$$I_1 I_3 \frac{d}{dt} \delta M_2 = (I_1 - I_3)\delta M_3 (\overline{M}_1 + \delta M_1),$$

$$I_2 I_1 \frac{d}{dt} \delta M_3 = (I_2 - I_1)(\overline{M}_1 + \delta M_1)\delta M_2,$$

so, for function V one has

$$\frac{dV}{dt} = \left\{ \frac{I_1 - I_3}{I_1 I_3}(\delta M_3)^2 + \frac{I_2 - I_1}{I_2 I_1}(\delta M_2)^2 \right\}(\overline{M}_1 + \delta M_1).$$

Choosing initial perturbations to have $V = V_0 > 0$ at $t = t_0$, one arrives at $dV/dt > 0$ (for $\overline{M}_1 + \delta M_1 > 0$). This means that a positive constant $L > 0$ can be found, such that $dV/dt \geq L$. That is why $V \geq V_0 + L(t - t_0)$, i.e., however small the initial value V_0 is taken, after a sufficiently long time-interval $t - t_0$, the magnitude of V will exceed any *a priori* given value of ε. The latter means instability (Chetaev, 1990).

After such an introduction, consider a two-dimensional barotropic model of the atmosphere over the spherical Earth, which is governed by the equation of conservation of absolute vorticity η, taken under quasi-solenoidal approximation (see Section 2.4)

$$\frac{\partial}{\partial t}\eta + J(\psi, \eta) = 0. \tag{4}$$

Here ψ is the stream function

$$\eta = \frac{1}{a^2 \sin\vartheta}\frac{\partial}{\partial\vartheta}\sin\vartheta\frac{\partial\psi}{\partial\vartheta} + \frac{1}{a^2 \sin^2\vartheta}\frac{\partial^2\psi}{\partial\lambda^2} + 2\Omega\cos\vartheta,$$

J denotes the Jacobian operator

$$J(A,B) = \frac{1}{a^2 \sin\vartheta}\left(\frac{\partial A}{\partial\vartheta}\frac{\partial B}{\partial\lambda} - \frac{\partial A}{\partial\lambda}\frac{\partial B}{\partial\vartheta}\right),$$

written for two arbitrary functions A and B; λ is the geographic longitude and ϑ is the co-latitude.

We proceed from the expression for the axial component of the atmospheric absolute angular momentum in terms of unit mass of an air column positioned over a unit area of the Earth's surface

$$M = \iint\left\{\frac{1}{a}\frac{\partial\psi}{\partial\vartheta} + a\Omega\sin\vartheta\right\}a\sin\vartheta\,d\sigma.$$

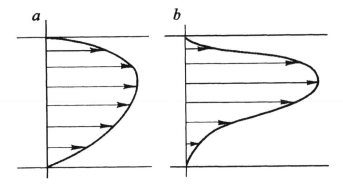

FIGURE 3   Examples of plane unidirectional flows for which Rayleigh's criterion on hydrodynamic stability holds (*a*) or does not hold (*b*), respectively.

Integrating by parts we find that

$$M = \iint \eta \cos \vartheta a^2 d\sigma.$$

This is an analog of the well-known two-dimensional fluid dynamics formulae for the Cartesian components of the fluid momentum (Batchelor, 1967). The conservation law

$$F = \iint \Phi(\eta) d\sigma,$$

where Φ is an arbitrary function, also holds for Equation (4).[1]

We seek the extremum of M provided that $F = $ const (cf. Fjortoft, 1950). For this, we find the absolute extremum of a functional $I = M + F$. The conditions for the extremum of I are the vanishing of the first variation $\delta I = 0$ and the definite sign of the second variation $\delta^2 I$. From the necessary extremum condition

$$\delta I = \iint \{\cos \vartheta a^2 + \Phi'(\eta)\} \delta\eta \, d\sigma = 0$$

it follows that the vorticity field η may be only zonal, if taken in an 'extremum point'. As it is seen from the formula for the second variation

---

[1] In modern literature, the latter constants of fluid motion are often referred to as *Casimir functionals* or simply *Casimirs* (cf. Salmon, 1998).

$$\delta^2 I = \iint \Phi''(\eta)(\delta\eta)^2 d\sigma = \iint \frac{a^2 \sin\vartheta}{d\eta/d\vartheta}(\delta\eta)^2 d\sigma,$$

the angular momentum M is extreme, if $\eta(\vartheta)$ is a monotonic function. This means stability after Lyapounov for such a zonal flow. In such a way, we arrive at Rayleigh–Kuo's criterion for barotropic stability (Kuo, 1949; Rayleigh, 1880; see also Diky, 1976).

In the case of no background fluid rotation ($\Omega = 0$), in order for the instability to occur it is necessary that at least one inflection point exists for the angular momentum density profile

$$m(x) = -(1 - x^2)(\partial\psi/\partial x),$$

where $x = \cos\vartheta$ and $2\pi a^2 m(x)dx$ is the angular momentum of an air ring enclosed between two latitudinal circles $x = $ const and $x + dx = $ const. For a plane, unidirectional shear flow, the stability criterion is formulated in terms of the lack of inflexion points in the velocity profile (Rayleigh's (1880) criterion, see also Arnol'd (1965) and Lin (1966)). Examples of velocity profiles for which the Rayleigh's criterion holds or does not hold, respectively, are shown in Fig. 3. In this context, we should note that the instability of a permanent ellipsoid rotation about a middle axis is an analog of the instability of a plane unidirectional shear flow with an inflexion point in the velocity profile.

When we study the stability of mean non-zonal flows, i.e., of a harmonic Rossby wave of a finite amplitude, which is an exact stationary solution of non-linear Equation (4) in an appropriate co-moving reference system, it is necessary to invoke additionally an energy integral

$$E = \iint \frac{1}{2}(\nabla\psi)^2 d\sigma$$

and to construct the Lyapounov function, starting from a linear combination of constants of motion $E + U$M$ + F$, where U is a Lagrange multiplier of velocity dimension.

Curiously, it is possible to suggest an alternative method for deriving the Rayleigh–Kuo's criterion if the absolute vorticity η is used as a latitudinal coordinate and the latitude of an isoscalar line $\eta = $ const is taken as a dependent variable. When doing this, it is convenient to pass to the variable $x = \cos\vartheta$. Now, the formula for the atmospheric angular momentum takes the form

$$M = \iint \eta x a^4 dx d\lambda = -a^4 \iint \eta d\left(\frac{1-x^2}{2}\right) d\lambda$$

$$= a^4 \iint \frac{1}{2}(1-x^2) d\eta d\lambda.$$

It is assumed that non-zonal perturbations are so small that the monotonicity of the absolute vorticity η with latitude, which is necessary to assume in order for such a variable transformation to be justified, is not violated anywhere. The following averaging operator along the isoscalar lines η = const is introduced:

$$\bar{A} = (1/2\pi) \int\limits_{0}^{2\pi} A d\lambda.$$

In the framework of a conservative problem, one has $\bar{x}$ = invariant. In fact, the quantity $2\pi a^2 (1-\bar{x})$ coincides exactly with the 'polar cap' area encircled with a closed contour η = const and it is constant because of the two-dimensional incompressibility of motion. Due to Schwartz' inequality $\int_0^{2\pi} x^2 d\lambda \geq \int_0^{2\pi} \bar{x}^2 d\lambda$ the angular momentum M reaches its maximally possible value M_{max} in the case of a zonal flow, when $x = \bar{x}$. That is why the difference

$$M_{max} - M = a^4 \iint \frac{x^2 - \bar{x}^2}{2} d\eta d\lambda \qquad (5)$$

is positive, which proves the stability after Lyapounov for the case of zonal flows with the monotonic dependence $\eta(x)$.

When the Earth's surface relief is taken into account, the atmospheric angular momentum is not a conservative property anymore. Instead, the total angular momentum of the coupled system 'atmosphere–solid Earth' is constant now. The non-constancy of the atmospheric angular momentum results in a specific type of instability called the orographic instability (Charney and Devore, 1979; Paegle, 1979; Källen, 1984; Galin and Kirichkov, 1985), which should be added to more well-known atmospheric instability types, namely the barotropic and baroclinic ones.

With account for orography, the absolute vorticity conservation law should be replaced by the principle of potential vorticity conservation. Under quasi-solenoidal approximation, when $c_0^2 \to \infty$ (see Section 2.4), the potential vorticity can be represented by a sum of absolute vorticity and a certain additional term, which does not include an explicit time-dependence and is uniquely determined by the orography height. In this case, within the accuracy of a non-significant additive constant, the atmospheric angular momentum is determined by the functional of the field of potential vorticity $\tilde{\Omega}$

$$M = a^2 \iint \tilde{\Omega} x d\sigma,$$

and the angular momentum deficit (5) becomes a real quantity, which may contribute to the atmospheric angular momentum increase, due to resonant non-linear interactions between Rossby waves and a zonal flow under the control of orography. This effect is discussed in Pisnichenko (1986), who also gives an approximate account for atmospheric large-scale compressibility.

The generalization of the above considerations for the case of a baroclinic atmosphere is not as straightforward as one may want. The main difficulty is associated with the necessity to account for essentially non-isentropic Earth's surface.

3.2 Baroclinic instability

Hereinafter, a more complicated problem of zonal flow stability within a three-dimensional baroclinic atmospheric model is considered. The beta-plane approximation is used, the air flow is assumed to be periodic in the zonal direction with a period L, the latter being equal to the latitudinal circle average length. On both its northern and southern sides, the flow is bounded by the vertical walls positioned strictly along the latitudinal circles.

Under quasi-geostrophic approximation and the neglect of both diabatic heating and frictional and other non-potential forces, atmospheric motion is governed by the non-linear equation of potential vorticity conservation (5) from Section 2.3

$$\frac{D}{Dt} q = 0, \quad q = \nabla^2 \psi + l + \frac{\partial}{\partial p} m^2 p^2 \frac{\partial \psi}{\partial p}. \tag{1}$$

It is convenient to consider the atmosphere to be bounded from above by the isobaric surface $p = \hat{p}$, where the following boundary condition is used

$$\frac{D}{Dt}\xi = 0, \quad \xi = -p\frac{\partial \psi}{\partial p}, \quad p = \hat{p}, \tag{1'}$$

the variable ξ being proportional to air temperature. As it has been already stressed in Section 2.3, the condition of vertical velocity vanishing at the Earth's surface Σ can be approximately replaced by the corresponding condition at a lower isobaric level $p = p_0$:

$$\frac{D}{Dt}\Theta = 0, \quad \Theta = -p\frac{\partial\psi}{\partial p} - \alpha^2\psi, \tag{1''}$$

where the variable Θ is proportional to surface air potential temperature. It is necessary to adopt some boundary conditions at the vertical walls. Consider these boundaries to be impermeable for air parcels

$$\partial\psi/\partial x = 0, \quad y = y_*, y^*,$$

and the velocity circulation to be time-constant at each baric level

$$\int (\partial\psi/\partial y)dx = \text{const}, \quad y = y_*, y^*.$$

The foregoing arguments are related to the existence of constants of motion. The first constant is energy, which is the sum of kinetic energy, baroclinic available potential energy and barotropic available potential energy

$$E = \frac{1}{2}\iiint\left\{(\nabla\psi)^2 + m^2\left(p\frac{\partial\psi}{\partial p}\right)^2\right\}dxdydp + \frac{1}{2}\iint_\Sigma m^2\alpha^2 p\psi^2 dxdy.$$

It is readily checked that E is constant, due to the governing equations

$$\frac{dE}{dt} = \iiint\{\nabla\psi\cdot\nabla\psi_t + m^2 p^2\psi_p\psi_{pt}\}\,dxdydp + \iint_\Sigma m^2\alpha^2 p\psi\psi_t dxdy$$

$$= -\iiint \psi q_t dxdydp - \iint_\Sigma m^2 p\psi\Theta_t dxdy + \iint m^2 p\psi\xi_t dxdy$$

$$= \iiint \psi J(\psi, q)dxdydp + \iint_\Sigma m^2 p\psi J(\psi,\Theta)dxdy$$

$$- \iint m^2 p\psi J(\psi,\xi)dxdy = 0.$$

Symbols $\iint\limits_{\Sigma}$ and $\iint$ denote integration over the isobaric surfaces $p = p_0$ and $p = \hat{p}$, respectively. In this section, subscripts x, y, p, t denote taking derivatives with respect to the corresponding variables. Transformation of both volume and surface integrals has been performed using integration by parts with the help of boundary conditions.

Secondly, the zonal component of atmospheric momentum is constant

$$P = -\iiint \frac{\partial\psi}{\partial y} dxdydp - \iint\limits_{\Sigma} ym^2\alpha^2 p\psi\,dxdy.$$

Actually,

$$\frac{dP}{dt} = -\iiint \psi_{yt}\,dxdydp - \iint\limits_{\Sigma} ym^2\alpha^2 p\psi_t\,dxdy$$

$$= -\iiint \left\{ \frac{\partial}{\partial y}\left(y\psi_{yt}\right) - y\psi_{yyt} - y\psi_{xxt} - y\frac{\partial}{\partial p}m^2p^2\frac{\partial\psi_t}{\partial p} \right\} dxdydp$$

$$- \iiint y\frac{\partial}{\partial}m^2p^2\frac{\partial\psi_t}{\partial} dxdydp - \iint\limits_{\Sigma} ym^2\alpha^2 p\psi\,dxdy$$

$$= \iiint yq_t\,dxdydp - \iint\limits_{\Sigma} m^2p^2 y\psi_{pt}\,dxdy + \iint m^2p^2 y\psi_{pt}\,dxdy$$

$$- \iint\limits_{\Sigma} m^2\alpha^2 py\psi_t\,dxdy = \iiint yq_t\,dxdydp$$

$$+ \iint\limits_{\Sigma} ym^2 p\Theta_t\,dxdy - \iint ym^2 p\xi_t\,dxdy.$$

We use Equations $(1-1'')$ in the following form

$$\zeta_t + \frac{\partial}{\partial y}\left(\psi_x\zeta\right) - \frac{\partial}{\partial x}\left(\psi_y\zeta\right) = 0, \quad \zeta = q,\Theta,\xi$$

and extend the chain of equalities further on

$$\frac{dP}{dt} = -\iiint y \frac{\partial}{\partial y}(\psi_x q) dxdydp - \iint_{\Sigma} ym^2 p \frac{\partial}{\partial y}(\psi_x \Theta) dxdy$$

$$+ \iint ym^2 p \frac{\partial}{\partial y}(\psi_x \xi) dxdy = \iiint \psi_x q \, dxdydp$$

$$+ \iint_{\Sigma} m^2 p \psi_x \Theta dxdy - \iint m^2 p \psi_x \xi dxdy = \iiint \{\psi_x \psi_{xx} + \psi_x \psi_{yy}$$

$$+ \psi_x l + \psi_x \frac{\partial}{\partial p} m^2 p^2 \frac{\partial \psi}{\partial p}\} dxdydp + \iint_{\Sigma} \psi_x m^2 p(-p\psi_p$$

$$- \alpha^2 \psi) dxdy - \iint \psi_x m^2 p(-p\psi_p) dxdy$$

$$= \iiint \{\frac{\partial}{\partial y}(\psi_x \psi_y) - \frac{\partial}{\partial x}(\frac{1}{2}\psi_y^2 + \frac{1}{2}m^2 p^2 \psi_p^2)\} dxdydp = 0,$$

which proves the statement declared.

Third, because the potential vorticity q is a material constant for the motion in isobaric surfaces and the area at every baric level $p = $ const is also constant, the following infinite set of integrals

$$F = \iiint \Phi(p,q) dxdydp$$

must be constant, where Φ is an arbitrary function of two variables. This could be also readily checked by formal calculations

$$\frac{dF}{dt} = \iiint \frac{\partial \Phi}{\partial q} q_t \, dxdydp = -\iiint \frac{\partial \Phi}{\partial q} J(\psi,q) dxdydp$$

$$= -\iiint J(\psi,\Phi) dxdydp = -\iiint \{\frac{\partial}{\partial y}(\psi_x \Phi)$$

$$- \frac{\partial}{\partial x}(\psi_y \Phi)\} dxdydp = 0.$$

Because of the boundary conditions, the following quantities

$$G = \iint_{\Sigma} \Gamma(\Theta)dxdy, \quad H = \iint X(\xi)dxdy$$

are constant, where Γ and X are arbitrary functions.

We compose a linear combination of constants of motion $I = E - UP + F + G + H$, where a constant quantity U has the dimension of velocity. Let ψ_0 be a given zonal flow, the stability of which is studied. The quantity $I[\psi] - I[\psi_0]$ is a constant of motion. Assuming a perturbation $\delta\psi = \psi - \psi_0$ be sufficiently small, we decompose $I[\psi] - I[\psi_0]$ onto series arranged according to the powers of $\delta\psi$ and its derivatives

$$I[\psi] - I[\psi_0] = \delta I + \frac{1}{2}\delta^2 I + ...,$$

i.e., onto the series of variations of increasing order. By the appropriate choice of arbitrary functions Φ, Γ, and X, one arrives at a vanishing first variation δI for any given zonal flow ψ_0 and for any arbitrarily taken constant quantity U. Let us carry out these computations

$$\delta I = \iiint \left\{ \nabla\psi \cdot \nabla\delta\psi + m^2 p^2 \psi_p \delta\psi_p + U\delta\psi_y + \frac{\partial\Phi}{\partial q}\delta q \right\} dxdydp$$

$$+ \iint_{\Sigma} \left\{ m^2\alpha^2 p\psi\delta\psi + Uym^2\alpha^2 p\delta\psi + \Gamma'\delta\Theta \right\} dxdy + \iint X'\delta\xi dxdy$$

$$= \iiint \left\{ -\psi - Uy + \partial\Phi/\partial q \right\}\delta q \, dxdydp$$

$$+ \iint_{\Sigma} \left\{ -m^2 p\psi - m^2 pUy + \Gamma' \right\}\delta\Theta \, dxdy$$

$$+ \iint \left\{ m^2 p\psi + m^2 pUy + X' \right\}\delta\xi \, dxdy.$$

Here, integration by parts has been used and, what is more, the $\delta\psi$-field has been assumed to satisfy the following conditions

$$\int \delta\psi_y dx = 0, \quad y = y_*, y^*,$$

which agree well with the boundary conditions. In order for the first variation to vanish, at the substitution $\psi = \psi_0$, it has to be

$$\left.\frac{\partial \Phi(p,q)}{\partial q}\right|_{q=q_0} = \psi_0 + Uy, \quad p \in (\hat{p}, p_0),$$

$$\left.\Gamma'(\Theta)\right|_{\Theta=\Theta_0} = m^2 p(\psi_0 + Uy), \quad p = p_0,$$

$$\left.X'(\xi)\right|_{\xi=\xi_0} = -m^2 p(\psi_0 + Uy), \quad p = \hat{p}. \tag{2}$$

If for every p-value we assume the monotonic behavior of q_0 with respect to latitude y, and the same property is assumed for both Θ at $p = p_0$ and ξ at $p = \hat{p}$, then conditions (2) can be satisfied.

Now, we calculate the second variation

$$\delta^2 I = \iiint \left\{ (\nabla \delta\psi)^2 + m^2 (p\delta\psi_p)^2 + \frac{\partial^2 \Phi}{\partial q^2} (\delta q)^2 \right\} dxdydp$$

$$+ \iint_\Sigma \left\{ m^2 \alpha^2 p(\delta\psi)^2 + \Gamma''(\delta\Theta)^2 \right\} dxdy + \iint X''(\delta\xi)^2 dxdy.$$

Using Equations (2), we shall have

$$\delta^2 I = \iiint \left\{ (\nabla \delta\psi)^2 + m^2 (p\delta\psi_p)^2 + \frac{\psi_{0y} + U}{q_{0y}} (\delta q)^2 \right\} dxdydp$$

$$+ \iint_\Sigma \left\{ m^2 \alpha^2 p(\delta\psi)^2 + m^2 p \frac{\psi_{0y} + U}{\Theta_{0y}} (\delta\Theta)^2 \right\} dxdy$$

$$- \iint m^2 p \frac{\psi_{0y} + U}{\xi_{0y}} (\delta\xi)^2 dxdy. \tag{3}$$

If the quadratic form (3) is positively defined, i.e., all coefficients before the field variables squared are positive, the zonal flow ψ_0 is stable after Lyapounov.

Indeed, consider a close neighborhood of the point $\psi = \psi_0$ in the functional space. Choosing it so small that the absolute value of a residual term in the decomposition of $I[\psi] - I[\psi_0]$ onto $\delta\psi$ powers becomes smaller than one-quarter of the second variation $\delta^2 I$, we arrive at the estimate

$$\frac{1}{4}\delta^2 I \le I[\psi] - I[\psi_0] \le \frac{3}{4}\delta^2 I.$$

If we use the sum of integrals

$$\|\delta\psi\|^2 = \iiint \left\{ (\nabla\delta\psi)^2 + m^2\left(p\delta\psi_p\right)^2 + m^{-2}\left(\delta q\right)^2 \right\} dxdydp$$

$$+ \iint_{\Sigma} \left\{ m^2\alpha^2 p(\delta\psi)^2 + m^2 p(\delta\Theta)^2 \right\} dxdy + \iint m^2 p(\delta\xi)^2 dxdy$$

as a norm in the functional space, i.e., as the measure of deviations of $\delta\psi$ from zero, then because of the positiveness of coefficients in (3), this norm will be equivalent to the second variation

$$c_1\|\delta\psi\|^2 \le \delta^2 I \le c_2\|\delta\psi\|^2,$$

with

$$c_1 = \min\left\{ m^2\frac{\psi_{0y} + U}{q_{0y}}, \quad \frac{\psi_{0y} + U}{\Theta_{0y}}, \quad -\frac{\psi_{0y} + U}{\xi_{0y}} \right\},$$

when this minimum is strictly less than unity and $c_1 = 1$ in the opposite case. In an analogous way,

$$c_2 = \max\left\{ m^2\frac{\psi_{0y} + U}{q_{0y}}, \quad \frac{\psi_{0y} + U}{\Theta_{0y}}, \quad -\frac{\psi_{0y} + U}{\xi_{0y}} \right\},$$

when this maximum is strictly greater than unity and $c_2 = 1$ in the opposite case. That is why

$$c_3\|\delta\psi\|^2 \le I[\psi] - I[\psi_0] \le c_4\|\delta\psi\|^2,$$

where $c_3 = (1/4)c_1$ and $c_4 = (3/4)c_2$. When it is chosen that $\|\delta\psi\|^2 \leq \delta$ at an initial time instant $t = t_0$, then $I[\psi] - I[\psi_0] \leq c_4\delta$. But $I[\psi] - I[\psi_0] = \mathrm{inv}$, and for any finite time-instant $t > t_0$

$$c_3\|\delta\psi\|^2 \leq I[\psi] - I[\psi_0] \leq c_4\delta$$

or

$$\|\delta\psi\|^2 \leq (c_4/c_3)\delta < 2(c_4/c_3)\delta.$$

It is sufficient to take $\delta = (1/2)(c_3/c_4)\varepsilon$, where $\varepsilon > 0$ is an arbitrarily small but finite number, in order to obtain $\|\delta\psi\|^2 < \varepsilon$. This means the stability after Lyapounov.

What does this criterion give in practice, i.e., what kind of zonal flows satisfy it? First, from the very beginning, we have supposed that potential vorticity is a monotonic function of latitude at every baric level. The most natural is to assume that this is a monotonically increasing function, i.e., q_0 behaves similarly to the planetary vorticity, namely to the Coriolis parameter l. The observed atmospheric zonal flows have predominantly eastward direction, i.e., $\partial\psi_0/\partial y < 0$. Choosing the constant U, which has been absolutely arbitrary up to now, to exceed the maximum of the modulus of a main zonal flow wind speed, we see that all volume integrals in (3) become positive. The situation with the integrals over the lower boundary is much worse. The first of them is positive. As far as the second integral is concerned, in our choice of U it appears negative, because Θ_0 decreases towards the poles, as a rule. The decrease of temperature towards the poles is a destabilizing factor, and this is quite natural. The integral taken over the upper atmospheric boundary is positive when the temperature decreases towards the poles and is negative in the opposite case, the latter being regularly observed in the stratosphere. However, when $\hat{p} \to 0$, the contribution of the latter integral becomes negligible.

Nevertheless, in which case one could guarantee the stability? If we consider a uniform surface air potential temperature field and assume that the upper boundary is isothermal, we could introduce an essential simplification into all the preceding arguments. Namely, instead of the boundary conditions $(1', 1'')$ it is possible to use simpler conditions of vanishing for both Θ-perturbations at $p = p_0$ and ξ-perturbations at $p = \hat{p}$. Then all the surface integrals in Equation (3) disappear along with the difficulties associated with them. In such a way, we arrive at Charney–Stern (1962) criterion for baroclinic instability.

Note that this criterion could be arrived at, by using, instead of I, a simpler functional $\tilde{I} = P + F + G + H$, having an arbitrary zonal flow ψ_0

as its stationary point with the unique requirement that q_0, Θ_0 and ξ_0 are the monotonic function of latitude. The second variation of this functional is given by the formula

$$\delta^2 \tilde{I} = -\iiint \frac{(\delta q)^2}{q_{0y}} dx\,dy\,dp - \iint_{\Sigma} m^2 p \frac{(\delta \Theta)^2}{\Theta_{0y}} dx\,dy$$

$$+ \iint m^2 p \frac{(\delta \xi)^2}{\xi_{0y}} dx\,dy.$$

For isentropic Earth's surface and isothermal upper baric level, or simply at $\hat{p} \to 0$, the surface integrals are dropped out and $\delta^2 \tilde{I}$ is negatively defined for the case of $\partial q_0 / \partial y > 0$. That is why, in this particular case, the total momentum of a zonal flow with a monotonic increase towards a pole of potential vorticity at all p-levels is maximum among all possible atmospheric states, having the same air parcel distribution on potential vorticity values for every isobaric surface. Instability of such a zonal circulation would be accompanied, with necessity, by a decrease in its total momentum, which is impossible, because the latter quantity is a constant of motion. Thus, such zonal circulation is obviously stable with respect to small but finite-amplitude perturbations.

When the second variation (3) has no definite sign, a zonal flow is said to be unstable with respect to the metrics $\|\delta\psi\|^2$. Consider a perturbation $\delta\psi$ obeying the requirements

$$\left\|\delta\psi\right\|^2 < \nu, \quad t \geq \tau_0 > t_0, \tag{4}$$

where τ_0 and $\nu > 0$ are constants and, what is more, the τ_0 values are sufficiently large, and the ν values are, in contrast, sufficiently small. To prove the instability of the ψ_0-solution, it is sufficient to discover in the functional space at least one trajectory which falls outside the domain defined by Equation (4) for increasingly small numerical values of an initial perturbation $\|\delta\psi\|^2$ at $t = t_0$. When seeking such a trajectory it is admissible to linearize the governing equations with respect to both perturbation, $\delta\psi$ itself and its spatial derivatives. The solution of the instability problem is thus reduced to the linear problem analysis. Within the framework of the linear problem, the existence of at least one solution infinitely growing in time guarantees instability.

Consider a case when latitudinal changes in the Coriolis parameter can be neglected. Here, no favorable latitudinal direction exists for the increase of potential vorticity. For simplicity, let us assume that the potential vorticity field is uniform at every isobaric level. Due to governing equations, when staying within the framework of adiabatic and frictionless approximation, it is clear that irrespective of the exact form of the zonal flow disturbance we would always arrive at the atmospheric state with an even potential vorticity distribution on p-surfaces. Therefore, we replace Equation (1) by the condition of q perturbations vanishing within the entire atmospheric volume. Now, the corresponding volume integral in (3) disappears. The zonal flow stability is guaranteed when such a constant U exists that (cf. Blumen, 1968)

$$\frac{\psi_{0y} + U}{\Theta_{0y}} > 0, \quad p = p_0; \quad -\frac{\psi_{0y} + U}{\xi_{0y}} > 0, \quad p = \hat{p}. \tag{5}$$

In the opposite case the instability is possible. Condition (5) can be satisfied if and only if Θ_0 and ξ_0 have the opposite directions of growth. An exception is the case when $\hat{p} \to 0$ and the corresponding integral contribution is increasingly small. When the upper boundary is essential and the direction of the increase of ξ_0 coincides with that of Θ_0 at the Earth's surface, the necessary conditions for instability are satisfied. The linear analysis of instability supports this conclusion.

3.3 Linear analysis (Eady model)

A mathematical treatment becomes the easiest, when we consider a horizontal atmospheric layer, enclosed between two adjacent isobaric surfaces ($p_0 - \hat{p} << (p_0 + \hat{p})/2$), and make an assumption analogous to that of the atmospheric air weak compressibility. Now, the metric multiplier $n^2 = m^2 p^2$ in the expression for potential vorticity q is admittable to be considered as a constant. To investigate the baroclinic instability mechanism in its pure form, consider a main zonal flow, with wind speed depending on altitude only, wind shear magnitude being the measure of atmospheric baroclinicity. When assuming that the zonal velocity is given by the function $\bar{u} = \bar{u}(p)$, superposing small non-zonal perturbations on it, and linearizing Equation (1) of Section 3.2 with respect to these perturbations, we shall have

$$\frac{\partial q'}{\partial t} + \bar{u} \frac{\partial q'}{\partial x} + \frac{\partial \psi'}{\partial x} \frac{\partial \bar{q}}{\partial y} = 0. \tag{1}$$

As the main flow stream function is given by the formula $\overline{\psi} = -\overline{u}(p)y + F(p)$, where F is an arbitrary function, the corresponding potential vorticity field reads as

$$\overline{q} = \frac{\partial^2 \overline{\psi}}{\partial y^2} + n^2 \frac{\partial^2 \overline{\psi}}{\partial p^2} + l = l - n^2 \frac{d^2 \overline{u}}{dp^2} y + n^2 F''.$$

In order to maximum simplify the problem, assume that $\overline{u} = \Lambda(P - p)$ where Λ and P are certain constant quantities. Finally, one gets $\overline{q} = l + n^2 F''(p)$, where the Coriolis parameter l is taken to be constant. In such a way, Equation (1) takes the form

$$\left(\frac{\partial}{\partial t} + \overline{u} \frac{\partial}{\partial x} \right) q' = 0, \quad q' = \nabla^2 \psi' + n^2 \frac{\partial^2 \psi'}{\partial p^2}.$$

This equation is supplied with the boundary conditions $w^* = 0$, $p = \hat{p}, \hat{p}_0$. The latter are taken under approximation of motions of a spatial scale much smaller than that of the planetary ones, when the atmospheric horizontal compressibility could be neglected. It means that when L is a characteristic horizontal scale and L_0 is the Obukhov's synoptic scale, $(L/L_0)^2 \ll 1$. In terms of a stream function, the boundary conditions linearized with respect to perturbations take the form (cf. Equations (1′,1″) of Section 3.2)

$$\left(\frac{\partial}{\partial t} + \overline{u} \frac{\partial}{\partial x} \right) \frac{\partial \psi'}{\partial p} + \frac{\partial \psi'}{\partial x} \frac{\partial^2 \overline{\psi}}{\partial y \partial p} = 0, \quad p = \hat{p}, p_0.$$

Assuming that perturbations are independent of the latitude y and using the thermal wind relation $\partial^2 \overline{\psi}/\partial y \partial p = -d\overline{u}/dp = \Lambda$, we arrive at the Eady's (1949) problem:

$$\left(\frac{\partial}{\partial t} + \overline{u} \frac{\partial}{\partial x} \right) \left(\frac{\partial^2 \psi'}{\partial x^2} + n^2 \frac{\partial^2 \psi'}{\partial p^2} \right) = 0,$$

$$\left(\frac{\partial}{\partial t} + \overline{u} \frac{\partial}{\partial x} \right) \frac{\partial \psi'}{\partial p} + \Lambda \frac{\partial \psi'}{\partial x} = 0, \quad p = \hat{p}, p_0,$$

$$\overline{u} = \Lambda(P - p).$$

A solution of this problem is searched as a superposition of normal modes $\psi' = \hat{\psi}(p)\exp\{ik(x - ct)\}$. The amplitude of perturbation $\hat{\psi}$ is complex, the wave number k is real, and the parameter c which is, generally speaking, complex, becomes an eigenvalue of the problem. The main flow is unstable if c possesses a positive imaginary part. However, because our problem is conservative (its non-linear prototype permits an energy conservation law), then if one has a certain eigenvalue c_1, it follows with necessity that there should be the corresponding complex conjugate eigenvalue $c_2 = c_1^*$. That is why, to prove instability, it is sufficient to discover a complex eigenvalue c within the spectrum of our problem. When substituting the given solution in our equations, we shall get

$$ik(\overline{u} - c)\left(-k^2\hat{\psi} + n^2\frac{d^2\hat{\psi}}{dp^2}\right) = 0,$$

$$ik(\overline{u}_0 - c)\frac{d\hat{\psi}}{dp} + ik\Lambda\hat{\psi} = 0, \quad p = p_0,$$

$$ik(\hat{\overline{u}} - c)\frac{d\hat{\psi}}{dp} + ik\Lambda\hat{\psi} = 0, \quad p = \hat{p}. \tag{2}$$

Dividing the first Equation (2) by $\overline{u}(p) - c$ and, thus, dropping away a continuous spectrum of our problem, we arrive at the equation

$$d^2\hat{\psi}/dp^2 - \lambda^2\hat{\psi} = 0, \quad \lambda^2 = k^2/n^2,$$

which has a general solution

$$\hat{\psi} = A\cosh\lambda p + B\sinh\lambda p.$$

With account for the two last Equations (2) we shall have a system of two homogeneous equations for two unknown variables A and B. Equalizing the corresponding determinant to zero, we obtain the characteristic equation for the determination of the eigenvalues

$$c^2 - c(\overline{u}_0 + \hat{\overline{u}}) + \overline{u}_0\hat{\overline{u}} + (\hat{\overline{u}} - \overline{u}_0)\Lambda\lambda^{-1}\coth\{\lambda(p_0 - \hat{p})\} - \Lambda^2\lambda^{-2} = 0.$$

The question of instability has been reduced to the investigation of the sign of the resulting quadratic equation discriminant

$$D = \left(\hat{\tilde{u}} - \bar{u}_0\right)^2 \left\{1 - 4\alpha^{-2}(\alpha \coth \alpha - 1)\right\}, \quad \alpha = \lambda\left(p_0 - \hat{p}\right).$$

Using the identity $\coth \alpha = (\tanh(\alpha/2) + \coth(\alpha/2))/2$, the discriminant is re-written in the form

$$D = \frac{4}{\alpha^2}\left(\hat{\tilde{u}} - \bar{u}_0\right)^2 \left\{\left(\frac{\alpha}{2} - \coth\frac{\alpha}{2}\right)\left(\frac{\alpha}{2} - \tanh\frac{\alpha}{2}\right)\right\}.$$

As $\alpha/2 \geq \tanh(\alpha/2)$, the value of α critical for instability is determined through the transcendental equation $\alpha_c/2 = \coth(\alpha_c/2)$, having an approximate root $\alpha_c \approx 2.4$. When $\alpha > \alpha_c$, the solution of our problem is a superposition of two neutral modes. For the case of $\alpha < \alpha_c$ there exist two complex-conjugate roots c, i.e., the instability has a long-wave character. An increment of the growing mode is equal to

$$k \operatorname{Im} c = \frac{k}{\alpha}\left(\hat{\tilde{u}} - \bar{u}_0\right)\left\{\left(\coth\frac{\alpha}{2} - \frac{\alpha}{2}\right)\left(\frac{\alpha}{2} - \tanh\frac{\alpha}{2}\right)\right\}^{\frac{1}{2}}.$$

Despite the fact that $\operatorname{Im} c(\alpha)$-value reaches its maximum for ultralong wavelengths, one has that the increment $k\operatorname{Im} c \to 0$ at $k \to 0$. Thus, the value $\alpha_m \approx 1.75$ should exist, which maximises this increment. The most unstable mode wavelength is determined by the formula

$$\lambda_m = \frac{2\pi}{k_m} = \frac{2\pi\left(p_0 - \hat{p}\right)}{n\alpha_m} = \frac{2\pi\left(p_0 - \hat{p}\right)}{m\alpha_m\left(\left(p_0 + \hat{p}\right)/2\right)}.$$

Assuming for an estimate that $2\left(p_0 - \hat{p}\right)/\left(p_0 + \hat{p}\right) = 1$, we get $\lambda = 2\pi/m\alpha_m \approx 3600$ km. One-quarter of this wavelength, equal to 900 km, is fairly close to the characteristic scale of 1000 km, which had been laid in the basis of the quasi-geostrophic approximation of atmospheric dynamic equations. With the help of the parameter values just taken, the maximum increment is given by the formula $(k \operatorname{Im} c)_m \approx 0.306m\left(\hat{\tilde{u}} - \bar{u}_0\right)$ which corresponds to e-folding time of $\tau_m \cong 3.8$ days for $\hat{\tilde{u}} - \bar{u}_0 = 10$ m·s^{-1}.

For the sake of completeness let us study the continuous spectrum of the problem described by Equations (2). Integration of the first Equation (2) over p gives

$$\int_{\hat{p}}^{p_0} (\bar{u} - c) \frac{d^2 \hat{\psi}}{dp^2} dp = \int_{\hat{p}}^{p_0} (\bar{u} - c) \frac{k^2}{n^2} \hat{\psi} dp.$$

The left-hand side of this equality is transformed by integrations by parts, with the help of both the second and third Equations (2). Now, we have

$$\frac{k^2}{n^2} \int_{\hat{p}}^{p_0} (\bar{u} - c) \hat{\psi} dp = 0.$$

There exists a continuum of solutions of this equation of the following type $\hat{\psi} = \delta(p - \tilde{p})$, $\tilde{p} \in (\hat{p}, p_0)$. Every such eigenfunction corresponds to a real eigenvalue $\tilde{c} = \bar{u}(\tilde{p})$. It means that the continuous spectrum does not contain any growing solution.

The Eady model appears to be of significant theoretical interest. There are several reasons for it. One of them is that the model allows a lot of improvements without losing its analytical nature, e.g., with the account of fluid compressibility, of variable static stability, of non-geostrophicity of motion (the Eady model in semi-geostrophic coordinates), etc. (Williams, 1974; Bell and White, 1988; Rotunno and Fontini, 1989). The Eady model can be used for the interpretation of laboratory experimental results obtained using non-uniformly heated rotating cylindrical setups (dishpans) (Lorenz, 1967; Hide and Mason, 1975). A principal physical shortcoming of the model is the disregard of the stabilizing action of the beta-effect on the longest spatial modes. Nevertheless, it is possible to construct a theory which would be more correct in this respect but still sufficiently simple, using a two-layer model proposed by Phillips (1951).

3.4   Two-layer Phillips' model

Consider a model, with atmospheric characteristics continuously varying with altitude being replaced by an atmosphere, essentially composed of two layers: the first extending from p_0 to $p_0/2$ and the second from $p_0/2$ to 0 ($\hat{p} = 0$). We replace the quasi-geostrophic potential vorticity conservation Equation (1) of Section 3.2 by the system of two equivalent equations:

$$\frac{D}{Dt}(\nabla^2 \psi + l) = \tilde{l} \frac{\partial w^*}{\partial p}, \tag{1}$$

$$\frac{D}{Dt} p^2 \frac{\partial \psi}{\partial p} + \tilde{l} m^{-2} w^* = 0, \tag{2}$$

where $w^* = dp/dt$ and $\tilde{l}$ is the mean Coriolis parameter. Equation (1) is written on isobaric surfaces 750 and 250 hPa, and Equation (2) on the 500 hPa surface, both using the finite-difference approximation of the derivatives with respect to pressure. The approximate boundary condition $w^* = 0$ is used at $p = 0$ and $p = p_0 = 1000$ hPa levels, which reliably describe the large-scale processes, with the only exception of planetary scale processes. The spatial scale of motions under consideration should be much less than the Obukhov's synoptic scale L_0. A typical profile $w^*(p)$ is characterized by an extremum in the vicinity of the atmospheric middle pressure level. For practical purposes, this profile could be often approximated by a parabolic function $w^* = W(p_0 - p)p$, where W depends both on horizontal co-ordinates and time. The fluid dynamic variables, taken at baric levels $p = 250$, 500 and 750 hPa, are supplied with indices '1', '3/2' and '2', respectively. As a result, one has:

$$\frac{D_1}{Dt}\left(\nabla^2 \psi_1 + l\right) = \tilde{l} \frac{w^*_{3/2}}{\left(p_0/2\right)}, \quad \frac{D_1}{Dt} = \frac{\partial}{\partial t} + J\left(\psi_1,\right),$$

$$\frac{D_2}{Dt}\left(\nabla^2 \psi_2 + l\right) = -\tilde{l} \frac{w^*_{3/2}}{\left(p_0/2\right)}, \quad \frac{D_2}{Dt} = \frac{\partial}{\partial t} + J\left(\psi_2,\right),$$

$$\frac{D_{3/2}}{Dt}\left(\frac{p_0}{2}\right)^2 \frac{p_2 - p_1}{\left(p_0/2\right)} + \tilde{l} m^{-2} w^*_{3/2} = 0, \quad \frac{D_{3/2}}{Dt} = \frac{\partial}{\partial t} + J\left(\psi_{3/2}\right).$$

To close the system, we set $\psi_{3/2} = (\psi_1 + \psi_2)/2$. Now, the following identities hold

$$\frac{D_1}{Dt}\left(\psi_1 - \psi_2\right) \equiv \frac{D_2}{Dt}\left(\psi_1 - \psi_2\right) \equiv \frac{D_{3/2}}{Dt}\left(\psi_1 - \psi_2\right).$$

When eliminating from the above written system the unknown $w^*_{3/2}$, we arrive at the atmospheric two-layer model equations (Phillips, 1951; Pedlosky, 1987)

$$\frac{\partial}{\partial t}\zeta_1 + J\left(\psi_1, \zeta_1\right) = 0, \quad \zeta_1 = \nabla^2 \psi_1 + l - m^2\left(\psi_1 - \psi_2\right), \tag{3}$$

$$\frac{\partial}{\partial t}\zeta_2 + J(\psi_2,\zeta_2)=0, \quad \zeta_2=\nabla^2\psi_2+l+m^2(\psi_1-\psi_2). \tag{4}$$

Writing the expression for potential vorticity q at 250 and 750 hPa levels, assuming that air temperature is finite at the top of the atmosphere, and using at the lower level $p = p_0$ so that

$$\Theta=-p\frac{\partial\psi}{\partial p}-\alpha^2\psi\approx-p\frac{\partial\psi}{\partial p},$$

it is easy to obtain the formula

$$q_1\approx\nabla^2\psi_1+l-m^2\left(\frac{p_0}{2}\right)^2\frac{\psi_1-\psi_2}{(p_0/2)^2}=\zeta_1.$$

At the same time,

$$q_2\approx\nabla^2\psi_2+l+m^2\left(\frac{p_0}{2}\right)^2\frac{\psi_1-\psi_2}{(p_0/2)^2}+m^2p_0^2\left.\frac{(\partial\psi/\partial p)}{p_0/2}\right|_{p=p_0}=\zeta_2-2m^2\Theta,$$

i.e., the invariant ζ_2 is a linear combination of q_2 and Θ.

The local Cartesian co-ordinates are used further on. As in the continuous model, the flow within a channel is studied with the boundary conditions on side walls, discussed in Section 3.2. Then, due to Equations (3, 4), the energy

$$E=\frac{1}{2}\iint\left\{(\nabla\psi_1)^2+(\nabla\psi_2)^2+m^2(\psi_1-\psi_2)^2\right\}dxdy,$$

momentum

$$P=-\iint\left(\frac{\partial\psi_1}{\partial y}+\frac{\partial\psi_2}{\partial y}\right)dxdy,$$

and an infinite set of integrals

$$F=\iint\left\{\Phi_1(\zeta_1)+\Phi_2(\zeta_2)\right\}dxdy,$$

with arbitrary functions Φ_1 and Φ_2, are constant. The method of checking these to be actually constants of motion is analogous to that performed in Section 3.2, for a continuously stratified atmosphere.

The stability of a zonal flow $\psi_1 = \overline{\psi_1}(y), \psi_2 = \overline{\psi_2}(y)$ is studied. This main flow is superimposed by small perturbations $\delta\psi_1$ and $\delta\psi_2$. A functional $I = E - UP + F$ is constructed, where U is an arbitrary constant quantity with the dimension of velocity. In order for the first variation of this functional

$$\delta I = \iint \left\{ \left(-\psi_1 - Uy + \Phi_1' \right)\delta\zeta_1 + \left(-\psi_2 - Uy + \Phi_2' \right)\delta\zeta_2 \right\}dxdy$$

to vanish through the substitution $\psi_1 = \overline{\psi_1}(y), \psi_2 = \overline{\psi_2}(y)$, it should be

$$\Phi_1'\left(\overline{\zeta_1} \right) = \overline{\psi_1} + Uy, \quad \Phi_2'\left(\overline{\zeta_2} \right) = \overline{\psi_2} + Uy. \tag{5}$$

If we assume the monotonicity of $\overline{\zeta_1}$ and $\overline{\zeta_2}$ with latitude, then these conditions could be always satisfied. Now, the second variation takes the form

$$\delta^2 I = \iint \left\{ \left(\nabla\delta\psi_1 \right)^2 + \left(\nabla\delta\psi_1 \right)^2 + m^2\left(\delta\psi_1 - \delta\psi_2 \right)^2 \right.$$

$$\left. + \frac{\left(d\overline{\psi_1}/dy \right) + U}{d\overline{\zeta_1}/dy}\left(\delta\zeta_1 \right)^2 + \frac{\left(d\overline{\psi_2}/dy \right) + U}{d\overline{\zeta_2}/dy}\left(\delta\zeta_2 \right)^2 \right\}dxdy. \tag{6}$$

The mean flow stability after Lyapounov is guaranteed when the integrand in (6) is positively defined for at least a single value of U. What do we have in reality? As in Section 3.2, it is natural to assume that $d\overline{\zeta_1}/dy > 0$. Thus, for all values of U which exceed the maximum of the modulus of zonal velocity at the 250 hPa level, the corresponding term in (6) is obviously positive. The problems arise with the lower level. We have shown that the invariant $\overline{\zeta_2}$ is a linear superposition of the potential vorticity $\overline{q_2}$ and surface potential temperature $\overline{\Theta}$. Moreover, according to the observational data, the terms $d\overline{q_2}/dy$ and $2m^2 d\overline{\Theta}/dy$ are close in absolute value but are of opposite sign. Thus, the derivative $d\overline{\zeta_2}/dy$ may vanish at certain latitudes. A necessary condition for instability is that the invariant $\overline{\zeta_2}$ fails to be a monotonically increasing function of latitude, contrary in this respect to $\overline{\zeta_1}$. It is possible to give a simple example of a flow, when this necessary condition for instability coincides with the sufficient condition.

Let us restrict ourselves with a class of flows with the zero horizontal wind shear, which means the case of pure baroclinic instability, as in Section 3.3. Now, one has $\overline{\psi}_1 = -\overline{u}_1 y + \text{const}$, $\overline{\psi}_2 = -\overline{u}_2 y + \text{const}$ and, what is more, $\overline{u}_1 > \overline{u}_2 > 0$. The critical case, separating stability and instability, corresponds to the condition

$$d\overline{\zeta}_2/dy = dl/dy - m^2(\overline{u}_1 - \overline{u}_2) = 0,$$

i.e., $\overline{u}_1 - \overline{u}_2 = m^{-2}(dl/dy)$ and, besides, when $\overline{u}_1 - \overline{u}_2 > m^{-2}(dl/dy)$, the necessary conditions for instability are satisfied. This estimate finds its confirmation in the results of a linear spectral theory, according to which the instability actually takes place when this inequality holds. A solution of Equations (3, 4), which are linearized with respect to $\delta\psi_1$ and $\delta\psi_2$, is searched in the form

$$\begin{Bmatrix} \delta\psi_1 \\ \delta\psi_2 \end{Bmatrix} = \begin{Bmatrix} A_1 \\ A_2 \end{Bmatrix} \exp\{ik_x(x - ct)\} \cos k_y y,$$

where A_1 and A_2 are the complex amplitudes, two real quantities k_x and k_y are the components of a wave vector and c is the eigen-, or spectral, parameter of the problem. When the latter is complex, one has instability, following analogous arguments in Section 3.3. Transparent enough algebraic manipulations, which could be found in (Thompson, 1961; Pedlosky, 1987), enable one to arrive at the explicit form of a neutral stability curve (see Figure 4):

$$(\overline{u}_1 - \overline{u}_2)^2 = \frac{4m^4(dl/dy)^2}{(k_x^2 + k_y^2)^2\left[4m^4 - (k_x^2 + k_y^2)^2\right]}. \tag{7}$$

A wave number for the most rapidly growing (most 'unstable') mode is given by the formula $\kappa = (k_x^2 + k_y^2)^{1/2} = m\sqrt[4]{2} \approx 1.19m$. This corresponds to the zonal wavelength $\lambda = 2\pi/m\sqrt[4]{2} \approx 6000\,\text{km}$. Here, the condition $k_y = 0$ is used, which accounts for the infinite channel extent in the meridional direction. As it follows from (7), there exists the lowest threshold for wind shear $\overline{u}_1 - \overline{u}_2 = m^{-2}(dl/dy) \approx 10\,\text{m·s}^{-1}$ leading to instability.

In the atmosphere over the Northern Hemisphere, in the latitudinal range of 30–60°, a fairly good agreement between zonal flow characteristics and the threshold value $\overline{u}_1 - \overline{u}_2$ derived for the two-layer model, has been

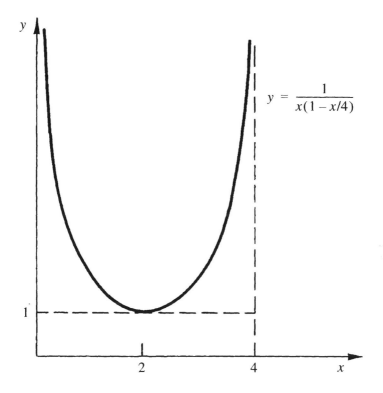

$$y = \frac{1}{x(1 - x/4)}$$

FIGURE 4   The neutral stability curve for the two-layer baroclinic model by Phillips (1951); $x = (k_x^2 + k_y^2)^2 m^{-4}$, $y = (\Delta u)^2 m^4 (dl/dy)^{-2}$.

established empirically (Stone, 1978). In the Southern Hemisphere, this correspondence is much worse, and the appearing asymmetry between the hemispheres needs to be explained, possibly by invoking the barotropic mechanisms of the loss of stability.

The Phillips' model is the simplest in an hierarchical set of multi-layer baroclinic atmospheric models which, when taken free from the quasi-geostrophic approximation, are laid in the basis of the modern numerical simulation of atmospheric general circulation and climate processes. For recently used models the number of levels approaches 30, or even more, which enables one to describe the fine peculiarities in the vertical atmospheric structure and to bring a model nearer to reality.

3.5   Vertical stability of atmospheric motions. Richardson's criterion

What is the physical mechanism of baroclinic instability studied in the preceding three sections? To answer this question, let us consider a simple

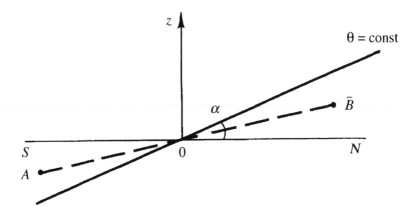

FIGURE 5 Disposition of isentropic surfaces accompanying the baro-clinic instability; α is the angle of slope of surfaces θ = const with respect to the horizontal.

scheme (see Figure 5), depicting the configuration of both isentropic and equipotential surfaces (the latter being the geopotential equiscalar surfaces) within a baroclinic atmosphere. A slope of isentropic surfaces with respect to a horizontal plane, characterized by a small angle $\alpha \approx 1/300$, is sustained by effects of both background rotation and vertical wind shear. The latter are necessary in order to establish a geostrophic balance at each horizontal level. When, for some reason, an air parcel makes an excursion from point O to either point A or point B, it has a chance to find itself in the ambience of air parcels with either larger (in point A) or, respectively, smaller (in point B) potential temperature values. An appearing buoyancy force will tend to turn the air parcel at a greater distance from its initial position. Thus, the baroclinic instability is a specific form of thermal convection and is determined by the mutual configuration of both isentropic and equipotential surfaces. For the Eady's model it could be strictly proven that the maximum rate of release of the available potential energy, which is stored in a main zonal flow, is accompanied by the excursion of a fluid parcel along the bisectrix of angle α.

Let us ignore the effects of both rotation and relation between the vertical wind shear and the horizontal slope of isentropic surfaces. It is clear that the vertical wind shear itself plays a destabilizing role because it is associated with a certain amount of kinetic energy available for the transformation into that of eddies. The stabilizing influence is due to the stable atmospheric density stratification, characterized by Brunt–Vaisala frequency $N = (gc_p^{-1}ds/dz)^{1/2}$. An appropriate measure of shear flow stability is the non-dimensional combination of field variables $Ri = N^2(\partial v/\partial z)^{-2}$, which is called the Richardson number. The smaller the Richardson number, the easier the instability arises. One should distinguish between three cases: (a) $Ri < 0$, i.e., $N^2 < 0$, when the convective instability perpetually happens, with the exclusion of very thin viscous and thermoconductive fluid layers, when the instability is determined by Rayleigh's (1916) criterion, (b) $Ri > Ri_{cr}$, when the flow is invariably stable, and (c) $0 < Ri < Ri_{cr}$, when the instability is possible.

What is the critical value Ri_{cr} of Richardson number? Both observations and various sorts of estimations show that $Ri_{cr} \leq 1$ (Monin and Yaglom, 1971; Miles, 1986). For large-scale atmospheric motions, one has, as a rule, $Ri \sim 10^2 >> Ri_{cr}$. Exceptions are sharp frontal zones which will be discussed at the very end of this section.

Richardson's criterion has a simple fluid dynamic meaning. Consider, for the convenience of the calculation, the case of an incompressible heterogeneous fluid, when $s = -c_p \ln\rho + \text{const}$ and $N = (-g d\ln\rho/dz)^{1/2}$. Within the fluid, we cut off a round cylinder with the horizontal axis. Let us assume that the mass of a cylinder's unit-length portion is equal to m and the corresponding momentum of inertia (with respect to the horizontal axis) is I. The cylinder's center of gravity lies below its geometrical center (to good accuracy, the center of inertia) at a distance of l_0. Deflecting the cylinder from its equilibrium state to a certain angle φ and/or conveying a certain initial angular velocity $\dot\varphi$ to it, we force the cylinder to oscillate. These oscillations are described by the governing equation of a physical pendulum

$$I\ddot\varphi + mgl_0 \sin\varphi = 0.$$

On assumption that at the initial time-instant $\dot\varphi = \omega$, $\varphi = 0$, after integration of this equation one gets

$$I\dot\varphi^2/2 + mgl_0(1 - \cos\varphi) = I\omega^2/2.$$

It is clear that in order to overturn the cylinder in the gravity field, it is necessary to satisfy the condition $mgl_0/I\omega^2 \leq 1/4$. It means that the kinetic energy of the initial revolution must be not smaller than the doubled potential energy deficit $2mgl_0$. It is not difficult to relate the parameters m, I, and l_0 to the Brunt–Vaisala frequency N. We use the fact that $l_0 = -(1/4)R^2(d\ln\rho/dz)$ and $I = 0.5mR^2$, where R is the cylinder's radius. As the result, we arrive at the criterion $N^2/2\omega^2 \leq 1/4$. If the kinetic energy of an initial push is taken from the kinetic energy stored in a unidirectional shear flow having a velocity profile $\bar{u}(z)$, then upon assumption that $\omega^2 = (d\bar{u}/dz)^2$, we formulate a condition for the shear flow instability as the condition of a 'fluid dynamic pendulum' overturning, in the form of criterion $N^2(d\bar{u}/dz)^{-2} \leq 1/2$, with $\mathrm{Ri}_{cr} = 1/2$. By its essence, this is the condition for wave-to-vortices transformation, when a fluid motion becomes turbulent and, thus, irreversible. That is why Richardson's criterion is often referred to as the condition for the non-decadence of turbulence.

Richardson's criterion, initially derived from energy balance considerations (see Brunt, 1941; Monin and Yaglom, 1971, 1975), has received an important support in results of the linear stability theory for stratified fluid unidirectional shear flows. Miles (1961) for a monotonic velocity profile $\bar{u}(z)$ and Howard (1961) for an arbitrary profile $\bar{u}(z)$ have shown that a sufficient condition for stability with respect to normal mode-like perturbations is the validity of inequality $\mathrm{Ri} > 1/4$ everywhere within a fluid flow domain. This is the Miles–Howard's criterion of stability. Nevertheless, in the framework of the non-linear theory, one is unable to formulate general conditions for stability after Lyapounov in the form of a universal Richardson's criterion, as it has been done for the Rayleigh–Kuo criterion for two-dimensional flows of a homogeneous fluid (Arnol'd, 1965, see Section 3.1) and for the Charney–Stern criterion in the case of quasi-geostrophic dynamics of a baroclinic atmosphere (Blumen, 1968; Diky and Kurgansky, 1971, see Section 3.2). Inevitably, instability with respect to short-wave perturbations is always permissible, the fact which has been pointed out in a pioneering paper by Diky (1965). Only when performing a finite-dimensional truncation of fluid dynamic equations one succeeds in formulating a sufficient stability condition in the Lyapounov sense. In such an approach, perturbations are assumed to have the minimum spatial scale λ. Surprisingly, nevertheless, the stability condition is written not in terms of Richardson's number but includes the combination of the velocity field and its second spatial derivative: $N^2(\bar{u}d^2\bar{u}/dz^2)^{-1}$ (Abarbanel et al., 1986; Kurgansky, 1988). A criterion $d^2(\bar{u}^2/2)/dz^2 < N^2$, different from the Richardson's criterion, appears in the analysis of the conditions for the existence of the stationary solutions of well-known Long's equation within the problem of stratified fluid shear flow over an obstacle (Blumen, 1989). Along with the increase in the order of the finite-dimensional dynamic system, which approximates the fluid dynamic equations, or, in other words, when decreasing the scale λ, the stability domain collapses and, in the framework of the infinite-dimensional problem, the instability is possible even when the corresponding linear problem analysis guarantees the stability with respect to normal modes, like in Miles (1961) and Howard (1961). The lack of complex eigenvalues in the linear problem spectrum does not obligatorily mean the stability after Lyapounov (Abarbanel et al., 1986).

This is a noticeable difficulty of the basic hydrodynamic stability theory. This issue contains a lot of problems and needs further investigations.

The Richardson's number appears in a natural way when treating the energetics of a stratified fluid mixing at the expense of the turbulence resulting from the hydrodynamic instability of a main flow with the horizontal velocity vector $\mathbf{v}(z)$. For a horizontally homogeneous fluid layer of depth h, with an infinite horizontal extent, the potential energy taken per fluid column of unit cross-section is equal to

$$\Pi = \int_{-h/2}^{h/2} gz\rho dz.$$

The origin of coordinates is placed in the midst of the layer. For the sake of simplicity, the case of an incompressible but heterogeneous fluid is considered. For a completely mixed layer the fluid density becomes height-independent, and, what is more, its constant value ρ_0 is determined from the mass conservation condition

$$m = \rho_0 h = \int_{-h/2}^{h/2} \rho dz.$$

The amount of work needed for mixing such a fluid is equal to the potential energy deficit

$$W = \int_{-h/2}^{h/2} gz\rho_0 dz - \int_{-h/2}^{h/2} gz\rho dz.$$

Let us restrict ourselves with a thin layer analysis and expand the $\rho(z)$-function to Taylor series with respect to altitude z:

$$\rho(z) = \rho(0) + \left.\frac{d\rho}{dz}\right|_{z=0} z + O(z^2).$$

For such a linear approximation, $\rho_0 = \rho(0)$, due to mass conservation. Here, the following formula holds (Turner, 1973; Ozmidov, 1983):

$$W \approx \frac{1}{12}\rho_0 N^2 h^3, \quad N^2 = -g(d\ln\rho/dz)\big|_{z=0}.$$

One would get just the same expression for a work for compressible hetero-geneous fluid mixing, if it is taken that $N^2 = gc_p^{-1}(ds/dz)$ (Kurgansky, 1983).

The work for fluid mixing within the layer can be performed at the expense of kinetic energy stored inside

$$K = \int_{-h/2}^{h/2} (1/2)\mathbf{v}^2 \rho \, dz.$$

However, not the total kinetic energy is actually available for the trans-formation into potential energy. The quantity K has the precise minimum, which is determined from the condition for the horizontal component of fluid motion momentum conservation

$$\mathbf{P} = \int_{-h/2}^{h/2} \mathbf{v} \rho \, dz.$$

Indeed, let us search for the minimum of K on condition that $\mathbf{P}$ and m are constants. We compose the functional $H = K + \lambda \cdot \mathbf{P} + \mu m$, where a constant vector λ and a constant scalar quantity μ play the role of non-defined Lagrange multipliers, and seek the absolute minimum of H. From the condition of the first variation δH vanishing we have $\mathbf{v} = -\lambda = \mathbf{v}_0 = $ const. Now, the minimum kinetic energy is equal to

$$K_{\min} = \int_{-h/2}^{h/2} (1/2)\mathbf{v}_0^2 \rho \, dz = \mathbf{P}^2/2m,$$

i.e., is completely determined by both the momentum and mass values. The difference $K' = K - K_{\min}$ is called the available kinetic energy (Starr, 1966). Expanding the horizontal velocity field $\mathbf{v}$ onto Taylor series with respect to z

$$\mathbf{v}(z) = \mathbf{v}(0) + \left(d\mathbf{v}/dz\right)\big|_{z=0} z + O(z^2)$$

and restricting ourselves to linear terms only, we shall get

$$K' = \frac{1}{24} \rho_0 \left[\left(d\mathbf{v}/dz\right)\big|_{z=0}\right]^2 h^3.$$

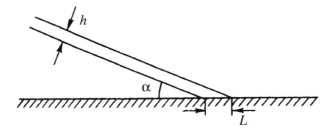

FIGURE 6   Sketch of a frontal layer separating two distinct air masses; h is the frontal layer thickness, α is the angle of slope of a frontal surface with respect to the horizontal, L is the width of a zone formed by the intersection between the frontal layer and the Earth's surface.

As the result, in order for fluid mixing to occur, the following condition should be satisfied:

$$\mathrm{Ri} = N^2 \left(d\mathbf{v}/dz \right)^{-2} \leq 1/2.$$

To overturn all the fluid within the layer the double amount of work, equal to twice as large potential energy deficit, is needed. Herewith, the condition $\mathrm{Ri} \leq 1/4$ must be satisfied.

As an illustration of the application of Richardson's criterion to large-scale atmospheric dynamic problems, let us estimate the physical width h of a frontal discontinuity, using a simplified theoretical model of atmospheric fronts proposed by Margules. Let us suppose that we have a plane stationary separation surface between two air masses with all thermodynamic parameters, except pressure, having a jump across this surface. To a good accuracy the following formula holds for an angle α of the slope of this surface with respect to the horizontal plane

$$\tan \alpha \approx \frac{l\theta}{g} \frac{\Delta u}{\Delta \theta}. \qquad (1)$$

Here $\Delta\theta/\theta$ is the relative potential temperature jump across the front, and Δu is the corresponding wind speed jump. The derivation of formula (1) is given in numerous texts on dynamic and synoptic meteorology (Brunt, 1941; Haltiner and Martin, 1957; Palmen and Newton, 1969). Here, we only note that this is the integral form of the thermal wind equation $\partial u/\partial z \approx -gl^{-1}(\partial\ln\theta/\partial z)$. The representation of the front by the separation surface is, certainly, a crude idealization. In reality, one should speak of a sharp transitional zone between two air masses, inside which the meteorological fields change abruptly. A rough estimate of such a transitional layer width could be performed, starting with the assumption that a self-sustained Richardson's number value, close to the critical value for the onset of instability, is established in the layer (cf. Brunt, 1941): $Ri \leq Ri_{cr}$. Indeed, assume that Ri becomes significantly less than Ri_{cr}. The vertical air turbulent mixing would immediately develop, which will smooth the spatial gradients in meteorological fields and correspondingly increase the value of Ri. If, in contrast, the value of Ri would become noticeably greater than unity, the large-scale advective (frontogenetic) processes, associated with non-zero deformation velocity fields, will play their role, thus reducing Ri up to its critical value. Replacing the vertical derivatives in the definition of Richardson's number by finite differences, we shall have

$$Ri = \frac{g\Delta\theta h^2}{\theta h(\Delta u)^2} = \frac{g\Delta\theta h}{\theta(\Delta u)^2} \leq 1,$$

which results in

$$h \leq \theta(\Delta u)^2 / g\Delta\theta.$$

Assuming that $\Delta\theta/\theta \cong 10^{-2}$, $\Delta u \cong 3$ m·s^{-1} and $g \approx 10$ m·s^{-2}, we arrive at the estimate $h \leq 100$ m. The width of a band formed by intersection between the transitional frontal zone and the Earth's surface, which is the frontal zone width in its common sense (see Figure 6), is estimated by the value

$$L = \frac{h}{\sin\alpha} \geq \frac{h}{\tan\alpha} \approx \frac{\Delta u}{l} \approx 3 \times 10^4 \, m,$$

which agrees with observations.

CHAPTER 4

Isentropic Analysis of Large-scale Processes

This chapter gives certain elements of a complex (using the potential vorticity field) isentropic analysis of large-scale atmospheric processes based on observational data handling. This is a development of the isentropic analysis well-known in meteorology (Namias, 1939; Rossby *et al.*, 1937; Bugaev, 1947). Material is presented largely based on the results of investigations carried out at the Institute of Atmospheric Physics, Russian Academy of Sciences, Moscow. The author was lucky to participate in this research activity. Hereafter, an attempt is made to give an account of theoretical fundamentals of the method in question. Section 4.3 stands a little bit aside and is devoted to the precise formulation of Lorenz (1955) available potential energy concept. This is done for two reasons. First, because of the necessity to complete Chapter 1 in the part which relates to atmospheric energetics. Second, certain concepts and ideas of Section 3 are essentially used further on in this chapter.

4.1 Isentropic coordinates

Describing large-scale atmospheric processes, a quasi-static approximation can be used with good accuracy. A corresponding estimate shows that the quasi-static approximation is appropriate for motions with horizontal spatial scale not less than 200 km. Dealing with quasi-static atmospheric dynamic equations, the use of isobaric, or p-coordinates is well-known to give major advantages. In these coordinates, pressure plays the role of a vertical coordinate, and the height of isobaric surfaces becomes a dependent variable (Eliassen, 1948). A system of p-coordinates, as well as a system of σ-coordinates derived from it, with the independent variable σ being equal to the ratio of pressure p to the surface pressure at a given geographic position (Phillips, 1957), is nowadays widely used for numerical weather forecasting and in atmospheric general-circulation and climate models.

However, from the theoretical viewpoint the isentropic coordinate system, with potential temperature θ as the altitude, seems to be more attractive (see Eliassen, 1987). Many theoretical results are expressed more clearly and precisely in the isentropic coordinates. In particular, under a quasi-static approximation, the potential vorticity conservation law is formulated in the simplest way in these coordinates, and it appears possible to give a precise mathematical formulation to the concept of available potential energy, which is important in meteorology. The above refers primarily to the processes which can be considered dry-adiabatic, but the advantages associated with the use of isentropic coordinates go beyond the framework of adiabatic approximation. In these coordinates the impact of diabatic heating can be estimated in the most correct and precise way.

It is assumed that potential temperature monotonically increases with altitude everywhere within the atmosphere, i.e., the latter is stable stratified. Transition from Cartesian (x, y, z) coordinates to θ-coordinates is performed using the formula

$$\varphi(x, z) = \varphi(x_\theta, \theta),$$

where φ is an arbitrary scalar function. For simplicity, we have fixed the values of the second spatial coordinate y and of the time t and provided the horizontal coordinates in θ-coordinate system by the subscript 'θ'. However, one should remember that x-coordinate is measured, as before, along the horizontal plane, so actually $x_\theta \equiv x$. Now, the equality for differentials is written in the form

$$\left(\frac{\partial \varphi}{\partial x}\right)_z dx + \frac{\partial \varphi}{\partial z} dz = \left(\frac{\partial \varphi}{\partial x}\right)_\theta dx_\theta + \frac{\partial \varphi}{\partial \theta} d\theta,$$

where a subscript 'z' or 'θ' denotes that the derivative is taken at a constant value of the corresponding variable. As

$$d\theta = \left(\frac{\partial \theta}{\partial x}\right)_z dx + \frac{\partial \theta}{\partial z} dz$$

and $dx_\theta \equiv dx$, it follows that

$$\left(\frac{\partial \varphi}{\partial x}\right)_z dx + \frac{\partial \varphi}{\partial z} dz = \left[\left(\frac{\partial \varphi}{\partial x}\right)_\theta + \frac{\partial \varphi}{\partial \theta}\left(\frac{\partial \theta}{\partial x}\right)_z\right] dx + \frac{\partial \varphi}{\partial \theta}\frac{\partial \theta}{\partial z} dz.$$

Equating the coefficients before independent differentials, we have

$$\left(\frac{\partial \varphi}{\partial x}\right)_z = \left(\frac{\partial \varphi}{\partial x}\right)_\theta + \frac{\partial \varphi}{\partial \theta}\left(\frac{\partial \theta}{\partial x}\right)_z, \quad \frac{\partial \varphi}{\partial z} = \frac{\partial \varphi}{\partial \theta}\frac{\partial \theta}{\partial z}.$$

In a special case of $\varphi = z$, it gives

$$0 = \left(\frac{\partial z}{\partial x}\right)_\theta + \frac{\partial z}{\partial \theta}\left(\frac{\partial \theta}{\partial x}\right)_z.$$

Combining this and the previous formula, one gets

$$\left(\frac{\partial \varphi}{\partial x}\right)_z = \left(\frac{\partial \varphi}{\partial x}\right)_\theta - \frac{\partial \varphi}{\partial \theta}\left(\frac{\partial z}{\partial \theta}\right)^{-1}\left(\frac{\partial z}{\partial x}\right)_\theta.$$

Analogous relations hold for the derivatives with respect to the second horizontal coordinate y and time t.

We write the equations of horizontal motion

$$\frac{Du}{Dt} - lv = -\frac{1}{\rho}\left(\frac{\partial p}{\partial x}\right)_z + F, \quad \frac{Dv}{Dt} + lu = -\frac{1}{\rho}\left(\frac{\partial p}{\partial y}\right)_z + G,$$

where F and G are the components of a non-potential force, including that of viscosity. With the account for the definition of potential temperature θ, these equations are easily re-written in terms of the Exner function $\Pi = c_p(p/p_{00})^k$, $k = R/c_p$:

$$\frac{Du}{Dt} - lv = -\theta\left(\frac{\partial \Pi}{\partial x}\right)_z + F, \quad \frac{Dv}{Dt} + lu = -\theta\left(\frac{\partial \Pi}{\partial y}\right)_z + G.$$

Here, the hydrostatic equation takes the form $\partial \Pi/\partial z = -g/\theta$. Performing a transition to θ-coordinates and following the above-written formulae taken at $\varphi = \Pi$,

$$\theta\left(\frac{\partial \Pi}{\partial x}\right)_z = \theta\left(\frac{\partial \Pi}{\partial x}\right)_\theta - \theta\frac{\partial \Pi}{\partial \theta}\left(\frac{\partial z}{\partial \theta}\right)^{-1}\left(\frac{\partial z}{\partial x}\right)_\theta,$$

$$= \left(\frac{\partial}{\partial x} (\theta \Pi) \right)_\theta - \theta \frac{\partial \Pi}{\partial z} \left(\frac{\partial z}{\partial x} \right)_\theta = \left(\frac{\partial}{\partial x} (\Pi \theta + gz) \right)_\theta,$$

one finally gets

$$\frac{Du}{Dt} - lv = - \left(\frac{\partial M}{\partial x} \right)_\theta + F, \quad \frac{Dv}{Dt} + lu = - \left(\frac{\partial M}{\partial y} \right)_\theta + G. \tag{1}$$

The function $M = \Pi \theta + gz \equiv c_p T + gz$ is called the Montgomery stream function, or the Montgomery isentropic potential, after a scientist who was the first to perform the transition to this variable (Montgomery, 1939). The Montgomery function is simply related to the Exner function by the formula $\partial M / \partial \theta = \Pi$, which replaces the hydrostatic equation in θ-coordinates.

The material time-derivative in θ-coordinates takes the form

$$\frac{D\varphi}{Dt} = \left(\frac{\partial \varphi}{\partial t} \right)_\theta + u \left(\frac{\partial \varphi}{\partial x} \right)_\theta + v \left(\frac{\partial \varphi}{\partial y} \right)_\theta + \dot\theta \frac{\partial \varphi}{\partial \theta}, \quad \dot\theta \equiv \frac{D\theta}{Dt},$$

φ being an arbitrary scalar function.

Thus, under adiabatic approximation, when $\dot\theta = 0$, air motion looks like a two-dimensional one. All fluid parcel trajectories lie in the isentropic surfaces. A 'three-dimensionality' appears only due to diabatic heating $\dot\theta = \hat{Q}/\Pi$, where $\hat{Q}$ are the specific heating rates per unit mass. Herewith, the trajectories of air parcels penetrate the isentropic surfaces, though in reality this penetration happens at very small angles, of the order of 10^{-4} (in radian measure), see Section 4.4.

In order to formulate a closed system of atmospheric dynamic equations in θ-coordinates, we have to re-write the mass-continuity equation

$$\frac{D\rho}{Dt} + \rho \left[\left(\frac{\partial u}{\partial x} \right)_z + \left(\frac{\partial v}{\partial y} \right)_z + \frac{\partial w}{\partial z} \right] = 0$$

in these coordinates. First, we have

$$\left(\frac{\partial u}{\partial x} \right)_z = \left(\frac{\partial u}{\partial x} \right)_\theta - \frac{\partial u}{\partial \theta} \left(\frac{\partial z}{\partial \theta} \right)^{-1} \left(\frac{\partial z}{\partial x} \right)_\theta,$$

$$\left(\frac{\partial v}{\partial y} \right)_z = \left(\frac{\partial v}{\partial y} \right)_\theta - \frac{\partial v}{\partial \theta} \left(\frac{\partial z}{\partial \theta} \right)^{-1} \left(\frac{\partial z}{\partial y} \right)_\theta.$$

Second, by definition $w = Dz/Dt$, and we can, consequently, write

$$\frac{\partial w}{\partial z} = \left(\frac{\partial z}{\partial \theta}\right)^{-1} \frac{\partial}{\partial \theta}\left[\left(\frac{\partial z}{\partial t}\right)_\theta + u\left(\frac{\partial z}{\partial x}\right)_\theta + v\left(\frac{\partial z}{\partial y}\right)_\theta + \dot\theta \frac{\partial z}{\partial \theta}\right]$$

$$= \left(\frac{\partial z}{\partial \theta}\right)^{-1}\left[\left(\frac{\partial}{\partial t}\left(\frac{\partial z}{\partial \theta}\right)\right)_\theta + u\left(\frac{\partial}{\partial x}\left(\frac{\partial z}{\partial \theta}\right)\right)_\theta + v\left(\frac{\partial}{\partial y}\left(\frac{\partial z}{\partial \theta}\right)\right)_\theta\right.$$

$$\left. + \dot\theta \frac{\partial}{\partial \theta}\left(\frac{\partial z}{\partial \theta}\right)\right] + \left(\frac{\partial z}{\partial \theta}\right)^{-1}\left(\frac{\partial u}{\partial \theta}\left(\frac{\partial z}{\partial x}\right)_\theta + \frac{\partial v}{\partial \theta}\left(\frac{\partial z}{\partial y}\right)_\theta + \frac{\partial \dot\theta}{\partial \theta}\frac{\partial z}{\partial \theta}\right).$$

Substituting these expressions in the mass-continuity equation and cancelling similar terms, we finally get

$$\frac{D}{Dt}\ln\left(\rho\frac{\partial z}{\partial \theta}\right) + \left(\frac{\partial u}{\partial x}\right)_\theta + \left(\frac{\partial v}{\partial y}\right)_\theta + \frac{\partial \dot\theta}{\partial \theta} = 0,$$

and, besides, $\rho\partial z/\partial \theta = -g^{-1}\partial p/\partial \theta$, due to the hydrostatic equation. The latter quantity will be denoted as ρ_θ later on.

We have to formulate the boundary condition at the Earth's surface. Here, it is necessary to mention the main difficulty, which accompanies the introduction of isentropic coordinates and has created until now a barrier for their wide use in the practise of synoptic analysis and numerical weather forecasting. The slope of isentropic surfaces to a horizontal plane is approximately 30 times as large as the corresponding slope for isobaric surfaces. Thus, the Earth's surface could not be considered an isentropic surface even approximately. In contrast, in p-coordinates the lower boundary condition is commonly written at an isobaric surface, which is the closest to the Earth's surface. In θ-coordinates the latter would be a sophisticated constructed surface, the exact form of which is determined implicitly by the solution of the problem. Assuming the absence of mountains and treating the air as an ideal fluid, we adopt the impermeability condition $w = 0$ for the Earth's surface $z = 0$. In isentropic coordinates this condition takes the form

$$\frac{D}{Dt}\left(\theta\frac{\partial M}{\partial \theta} - M\right) = 0, \qquad \theta\frac{\partial M}{\partial \theta} - M = 0.$$

An additional complication is provided by the fact that the atmosphere is, as a rule, unstably stratified just above the Earth's surface. Thus, to arrive at a surface potential temperature value, one either takes the θ-value at the very top of the atmospheric planetary boundary layer, approximately at 925 hPa, or extrapolates the θ-values from the upper levels beneath.

4.2 Theorem on potential vorticity

Ertel (1942) has shown that a quantity $I = \rho^{-1}(\nabla \times \mathbf{v} + 2\vec{\Omega}) \cdot \nabla \theta$, called the potential vorticity, is a material constant for adiabatic air motion occurring in the field of potential forces only. Under quasi-static approximation, a mathematical expression for Ertel's potential vorticity is as follows

$$I = \rho^{-1}\left\{ \left(\frac{\partial v}{\partial x} - \frac{\partial u}{\partial y} + l \right) \frac{\partial \theta}{\partial z} + \frac{\partial u}{\partial z}\frac{\partial \theta}{\partial y} - \frac{\partial v}{\partial z}\frac{\partial \theta}{\partial x} \right\}.$$

In isobaric coordinates one has, correspondingly,

$$I = g\left\{ \left[\left(\frac{\partial v}{\partial x} \right)_p - \left(\frac{\partial u}{\partial y} \right)_p + l \right]\left(-\frac{\partial \theta}{\partial p} \right) + \frac{\partial v}{\partial p}\left(\frac{\partial \theta}{\partial x} \right)_p - \frac{\partial u}{\partial p}\left(\frac{\partial \theta}{\partial y} \right)_p \right\},$$

where the subscript 'p' indicates that the derivatives are taken at constant pressure values. In braces, the first term describes the impact of the barotropic factors; the second and the third terms, the contribution of baroclinic ones. In extratropics, for large-scale processes, the contribution of the last two terms is fairly small. Using the thermal wind relation, it could be shown that this contribution is always negative and is equal by its relative value to the reciprocal of the Richardson's number Ri, the latter being as large as Ri ~ 10^2 for large-scale processes. That is why it is admissible to use an approximate formula

$$I \approx g\left[\left(\frac{\partial v}{\partial x} \right)_p - \left(\frac{\partial u}{\partial y} \right)_p + l \right]\left(-\frac{\partial \theta}{\partial p} \right) \tag{1}$$

in practical computations.

A certain advantage of the use of isentropic coordinates is that one gets a single-termed expression for I, which looks similar to Equation (1) but, however, combines both barotropic and baroclinic effects and thus appears to be exact in the framework of quasi-static approximation. Let us perform the necessary computations.

We eliminate the Montgomery function M from the horizontal motion equations (1) of Section 4.1 by their cross-differentiation and then subtracting one from the other:

$$\left(\frac{\partial}{\partial x}\left(\frac{Dv}{Dt}+lu\right)\right)_\theta - \left(\frac{\partial}{\partial y}\left(\frac{Du}{Dt}-lv\right)\right)_\theta = \left(\frac{\partial G}{\partial x}\right)_\theta - \left(\frac{\partial F}{\partial y}\right)_\theta.$$

Performing some simple identical transformations, one arrives at the following equation (Haynes and McIntyre, 1987)

$$\left(\frac{\partial}{\partial t}\omega_{a\theta}\right)_\theta + \left(\frac{\partial}{\partial x}\left(u\omega_{a\theta}+\dot\theta\frac{\partial v}{\partial\theta}-G\right)\right)_\theta$$

$$+\left(\frac{\partial}{\partial y}\left(v\omega_{a\theta}-\dot\theta\frac{\partial u}{\partial\theta}+F\right)\right)_\theta = 0, \tag{2}$$

where $\omega_{a\theta} = (\partial v/\partial x)_\theta - (\partial u/\partial y)_\theta + l$ is the absolute vorticity component orthogonal to isentropic surfaces. In such a way, we obtain a theorem on the conservation of the vorticity charge $\omega_{a\theta}$ in isentropic coordinates. An elegant method to derive Equation (2) is to start from Equations (1) of Section 4.1 taken in the Gromeka–Lamb form

$$\left(\frac{\partial u}{\partial t}\right)_\theta = -\left(\frac{\partial B}{\partial x}\right)_\theta + \omega_{a\theta}v - \dot\theta\frac{\partial u}{\partial\theta} + F,$$

$$\left(\frac{\partial v}{\partial t}\right)_\theta = -\left(\frac{\partial B}{\partial y}\right)_\theta - \omega_{a\theta}u - \dot\theta\frac{\partial v}{\partial\theta} + G,$$

$$B = \left(u^2+v^2\right)/2 + M.$$

By eliminating the Bernoulli function B from these equations, one immediately arrives at Equation (2). Note that there is no a term like $\partial\left(\dot\theta\omega_{a\theta}+...\right)/\partial\theta$ in the left-hand side of Equation (2). It indicates the absence of the diffusion of the vorticity charge across isentropic surfaces. From the mass-continuity equation,

$$\left(\frac{\partial\rho_\theta}{\partial t}\right)_\theta + \left(\frac{\partial}{\partial x}(u\rho_\theta)\right)_\theta + \left(\frac{\partial}{\partial y}(v\rho_\theta)\right)_\theta + \frac{\partial}{\partial\theta}(\dot\theta\rho_\theta) = 0, \tag{3}$$

it follows that the isentropic surfaces are permeable for air parcels, in so far as one speaks of diabatic heating. In the course of diabatic heating, isentropic surfaces descend through the air; and the other way round, when the air is cooling they lift.

Under adiabatic approximation ($\dot{\theta} = 0$) and the assumption of the potential character of external forces ($F = G = 0$), Equation (2) could be re-written in the form

$$\left(\frac{\partial}{\partial t}\rho_\theta\frac{\omega_{a\theta}}{\rho_\theta}\right)_\theta + \left(\frac{\partial}{\partial x}\left(u\rho_\theta\frac{\omega_{a\theta}}{\rho_\theta}\right)\right)_\theta + \left(\frac{\partial}{\partial y}\left(v\rho_\theta\frac{\omega_{a\theta}}{\rho_\theta}\right)\right)_\theta = 0.$$

On these grounds, with the account of Equation (3) taken under adiabatic approximation, one arrives at the theorem of potential vorticity conservation

$$DI/Dt = 0, \quad I = \omega_{a\theta}/\rho_\theta.$$

In a very similar form, this result was established by Rossby (1940), well before the most general Ertel's theorem appeared.

4.3 Precise formulation of the available potential energy concept

Isentropic coordinates are especially convenient for the general theoretical treatment of the problems related to atmospheric energetics. Let us formulate the energy conservation law in θ-coordinates. We start from multiplying Equation (1) of Section 4.1 by $\rho_\theta u$ and $\rho_\theta v$, correspondingly, and by summing up the results obtained

$$\rho_\theta\frac{D}{Dt}\frac{u^2+v^2}{2} = -\rho_\theta\left(u\frac{\partial M}{\partial x}+v\frac{\partial M}{\partial y}\right)+\rho_\theta(uF+vG)$$

$$= -\frac{\partial}{\partial x}(\rho_\theta uM)-\frac{\partial}{\partial y}(\rho_\theta vM)+M\left[\frac{\partial}{\partial x}(\rho_\theta u)+\frac{\partial}{\partial y}(\rho_\theta v)\right]$$

$$+\rho_\theta(uF+vG).$$

Hereinafter, the subscript 'θ' near the derivatives is omitted everywhere. Using the mass-continuity equation

$$\frac{\partial}{\partial x}(\rho_\theta u) + \frac{\partial}{\partial y}(\rho_\theta v) = -\frac{\partial}{\partial t}\rho_\theta - \frac{\partial}{\partial \theta}(\rho_\theta \dot{\theta}),$$

we have

$$\rho_\theta \frac{D}{Dt}\frac{u^2 + v^2}{2} = -\frac{\partial}{\partial x}(\rho_\theta u M) - \frac{\partial}{\partial y}(\rho_\theta v M)$$

$$-M\frac{\partial}{\partial t}\rho_\theta - M\frac{\partial}{\partial \theta}(\rho_\theta \dot{\theta}) + \rho_\theta(uF + vG).$$

Taking into account that $\rho_\theta = -g^{-1}\partial p/\partial\theta$, by definition, we transform the third term in the right-hand side of the previous equation

$$-M\frac{\partial}{\partial t}\rho_\theta = \frac{\partial}{\partial \theta}\left(g^{-1}M\frac{\partial p}{\partial t}\right) - g^{-1}\frac{\partial M}{\partial \theta}\frac{\partial p}{\partial t}.$$

Due to the hydrostatic equation $\partial M/\partial\theta = \Pi$, we have

$$-M\frac{\partial}{\partial t}\rho_\theta = \frac{\partial}{\partial \theta}\left(g^{-1}M\frac{\partial p}{\partial t}\right) - g^{-1}c_p\left(\frac{p}{p_{00}}\right)^k\frac{\partial p}{\partial t}$$

$$= \frac{\partial}{\partial \theta}\left(g^{-1}M\frac{\partial p}{\partial t}\right) - \frac{\partial}{\partial t}\left[c_p g^{-1}\frac{p^{k+1}}{(k+1)p_{00}^k}\right].$$

We transform the fourth term in the right-hand side of the same equation, as above

$$-M\frac{\partial}{\partial \theta}(\rho_\theta \dot{\theta}) = -\frac{\partial}{\partial \theta}(\rho_\theta \dot{\theta}M) + \frac{\partial M}{\partial \theta}\rho_\theta \dot{\theta} = -\frac{\partial}{\partial \theta}(\rho_\theta \dot{\theta}M) + \Pi\rho_\theta \dot{\theta}.$$

As the result, we arrive at the equation

$$\rho_\theta \frac{D}{Dt}\frac{u^2 + v^2}{2} = -\frac{\partial}{\partial x}(\rho_\theta u M) - \frac{\partial}{\partial y}(\rho_\theta v M) + \rho_\theta(uF + vG)$$

$$+ \Pi \dot{\theta} \rho_\theta + \frac{\partial}{\partial \theta} \left(g^{-1} M \frac{\partial p}{\partial t} - \rho_\theta \dot{\theta} M \right) - \frac{\partial}{\partial t} \left[c_p g^{-1} \frac{p^{k+1}}{(k+1)p_{00}^k} \right].$$

It remains to use the mass-continuity equation multiplied by $(u^2 + v^2)/2$, and to add the resulting equation to the one written above in order to obtain the total atmospheric energy balance equation:

$$\frac{\partial}{\partial t} \left(\rho_\theta \frac{u^2 + v^2}{2} + c_p g^{-1} \frac{p^{k+1}}{(k+1)p_{00}^k} \right) + \frac{\partial}{\partial x} \left[\rho_\theta u \left(\frac{u^2 + v^2}{2} + M \right) \right]$$

$$+ \frac{\partial}{\partial y} \left[\rho_\theta v \left(\frac{u^2 + v^2}{2} + M \right) \right] + \frac{\partial}{\partial \theta} \left[\rho_\theta \dot{\theta} \left(\frac{u^2 + v^2}{2} + M \right) \right.$$

$$\left. - g^{-1} M \frac{\partial p}{\partial t} \right] = \rho_\theta \left(uF + vG + \dot{\theta} \Pi \right). \tag{1}$$

As it was mentioned above, the use of the isentropic coordinates leads to a conceptual difficulty associated with essentially non-isentropic Earth's surface. If we consider the general problems of atmospheric energetics, this difficulty is overcome with the help of an artificial approach suggested by Lorenz (1955). Namely, let us assume that at a certain point on the Earth's surface with coordinates (x_0, y_0) we have $\theta = \theta_0$, $p = p_0$. Let us extrapolate the pressure field beneath the Earth's surface, assuming that $p = p_0$ at $x = x_0$, $y = y_0$, $0 \le \theta \le \theta_0$. In other words, isentropic surfaces are continuously prolonged beneath the Earth's surface, but the air underneath is considered to be weightless. At the Earth's surface, $z = 0$, by definition

$$M = \Pi \theta + gz = \Pi_0 \theta_0, \quad \Pi_0 = c_p \left(p_0 / p_{00} \right)^k. \tag{2}$$

Beneath the Earth's surface the hydrostatic equation $\partial M / \partial \theta = \Pi_0$ is also fulfilled. Now, for the range of $0 \le \theta \le \theta_0$ it follows that $M = \Pi_0 \theta$, where we assumed that $M = 0$ at $\theta = 0$. Only this expression for M is consistent with formula (2).

We integrate Equation (1) over the entire atmospheric volume, taking into account that $\rho_\theta \equiv 0$ beneath the Earth's surface, and $M = 0$ at $\theta = 0$:

$$\frac{d}{dt} \left\{ \iiint_{\theta_0}^{\infty} \rho_\theta \frac{u^2 + v^2}{2} d\theta dx dy + \iiint_{0}^{\infty} c_p g^{-1} \frac{p^{k+1}}{(k+1)p_{00}^k} d\theta dx dy \right\}$$

$$= \iint \int_{\theta_0}^{\infty} \rho_\theta \left(uF + vG + \dot{\theta}\Pi \right) d\theta dx dy . \qquad (3)$$

When integration in Equation (3) is extended only over a certain portion of the atmosphere, appropriate boundary conditions are imposed on the lateral boundaries of the integration domain. For example, it is sufficient either to restrict oneself to the impermeability condition or to assume a spatial periodicity with respect to one or both horizontal directions. At the top of the atmosphere, where $\theta \to \infty$, we assume that

$$\rho_\theta \dot{\theta}\left(\frac{u^2 + v^2}{2} + M \right) - g^{-1} M \frac{\partial p}{\partial t} \to 0.$$

Under quasi-static approximation in question, the total energy is represented as a sum of the kinetic energy of the horizontal component of motion and the total potential energy, which reads as follows

$$\iint \int_0^{\infty} c_p g^{-1} \frac{p^{k+1}}{(k+1)p_{00}^k} d\theta \, dx dy. \qquad (4)$$

Only a small fraction of the latter, related to the buoyancy forces, could be transformed into the kinetic energy of atmospheric motions in purely adiabatic processes. The maximal part of the potential energy capable of such a transition, is called the available potential energy. Let us find it.

Before doing it, we shall prove a lemma, which is necessary for further discussion (Dutton and Johnson, 1967). We integrate the mass-continuity equation

$$\frac{\partial}{\partial t}\rho_\theta + \frac{\partial}{\partial x}\left(\rho_\theta u\right) + \frac{\partial}{\partial y}\left(\rho_\theta v\right) + \frac{\partial}{\partial \theta}\left(\rho_\theta \dot{\theta}\right) = 0$$

over the atmospheric volume above an arbitrary but fixed isentropic surface, $\theta = $ const. With the account of the pressure field extrapolated beneath the Earth's surface, we get

$$\frac{\partial}{\partial t}\iint g^{-1} p(x, y, \theta, t) dx dy - \iint \rho_\theta \dot{\theta} dx dy$$

$$= -\iiint \left\{ \frac{\partial}{\partial x}\left(\rho_\theta u\right) + \frac{\partial}{\partial y}\left(\rho_\theta v\right) \right\} d\theta dx dy.$$

At the atmospheric upper boundary it is assumed that $\rho_\theta\dot\theta \to 0$. In the left-hand side, the integration is extended over the surface, $\theta = $ const. Using the impermeability condition at the lateral boundaries, we have

$$\frac{\partial}{\partial t}\iint g^{-1}p\,dx\,dy = \iint \rho_\theta\dot\theta\,dx\,dy.$$

We define the pressure $\tilde p$ averaged over isentropic surfaces, $\theta = $ const, by the formula $\tilde p\iint dx\,dy = \iint p\,dx\,dy$. Under adiabatic approximation the quantity $\tilde p = \tilde p(\theta)$ is a constant of motion; in the diabatic case $\tilde p = \tilde p(\theta,t)$ and

$$\frac{d}{dt}\iint\int_0^\infty c_p g^{-1}\frac{\tilde p^{k+1}}{(k+1)p_{00}^k}\,d\theta\,dx\,dy = \iint\int_{\theta_0}^\infty \tilde\Pi\rho_\theta\dot\theta\,d\theta\,dx\,dy, \qquad (5)$$

where $\tilde\Pi = c_p\left(\tilde p/p_{00}\right)^k$.

Subtracting Equation (5) from Equation (3), we finally have

$$\frac{d}{dt}\left\{\iint\int_{\theta_0}^\infty \rho_\theta\frac{u^2+v^2}{2}\,d\theta\,dx\,dy + \iint\int_0^\infty c_p g^{-1}\frac{p^{k+1}-\tilde p^{k+1}}{(k+1)p_{00}^k}\,d\theta\,dx\,dy\right\}$$

$$= \iint\int_{\theta_0}^\infty \left[uF + vG + \dot\theta\Pi(1 - \tilde\Pi/\Pi)\right]\rho_\theta\,d\theta\,dx\,dy. \qquad (6)$$

As it follows from the Goelder's inequality, the total potential energy (4) is minimal in the atmospheric state, when pressure p on isentropic surfaces is uniform and equal to the corresponding average pressure value $\tilde p$. This minimum is exactly equal to

$$\iint\int_0^\infty c_p g^{-1}\frac{\tilde p^{k+1}}{(k+1)p_{00}^k}\,d\theta\,dx\,dy. \qquad (7)$$

Thus, in the left-hand side of Equation (6) one observes a positively defined quantity, which is just the available potential energy after Lorenz (1955). The latter quantity, to an accuracy of the third order magnitude terms, is equal to the integral

$$A = \frac{1}{2} \iint \int_0^\infty c_p g^{-1} k \frac{\tilde{p}^{k+1}}{p_{00}^k} \left(\frac{p - \tilde{p}}{\tilde{p}} \right)^2 d\theta \, dxdy.$$

In fact, the exact formulation of the concept of available potential energy has not been called for in papers on atmospheric energetics, as far as we know. Most likely, it happened due to the difficulties associated with the practical usage of θ-coordinates, because the Earth's surface is non-isentropic. Usually, certain approximations of this exact concept are used. The first such an approximation was suggested by Lorenz (1955) himself:

$$A \approx \frac{1}{2} \iint \int_0^{p_0} \frac{\overline{T}}{\gamma_a - \gamma} \left(\frac{T - \overline{T}}{\overline{T}} \right)^2 dpdxdy. \tag{8}$$

Here, a bar above the variables denotes averaging over isobaric surfaces; the isobaric surface $p = p_0$ bounds the atmosphere from below. The quantity (8) enters, as a constituent, into the exact constant of motion for linearized (with respect to deviations from the reference atmospheric state) dynamic equations, as well as for a certain class of simplified non-linear atmospheric models, particularly, for the quasi-geostrophic model (see Van Mieghem, 1973).

Equation (8) is laid in the basis of currently available numerous estimates of the amount of available potential energy stored in the atmosphere, as well as of characteristics of the main energetic cycle in the atmosphere (Lorenz, 1967; Van Mieghem, 1973; Peixoto and Oort, 1992). Most famous are the estimations by Oort (1964), according to which one has $A = 55 \times 10^5$ J·m^{-2}, per unit Earth's surface area. It exceeds the kinetic energy $K = 15 \times 10^5$ J·m^{-2} more than three times. The generation of available potential energy within the atmospheric general circulation processes (see Equation (6))

$$\iint \int_{\theta_0}^\infty \dot{\theta} \Pi (1 - \overline{\Pi}/\Pi) \rho_\theta d\theta dxdy \approx \iint \int_0^{p_0} g^{-1} \hat{Q} (1 - \overline{T}/T) dpdxdy$$

is estimated by the value of 2.3 W·m^{-2} taken per unit area. Just at the same rate the available potential energy converts into the kinetic energy, mainly due to baroclinic instability, which is a special sort of a sloping thermal convection in the Coriolis force field. Afterwards, at just the same rate the kinetic energy dissipates into heat, the major part of the latter going for the increase of the unavailable potential energy (7). The structural diagram of it is shown in Figure 7.

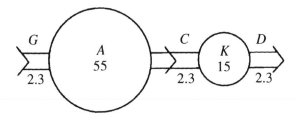

FIGURE 7   Main atmospheric energy cycle, after Oort (1964). Available potential energy A and kinetic energy K values are given in 10^5 J·m^{-2} units, theirs conversion rate C and dissipation rate D are expressed in W·m^{-2} units.

4.4   Diabatic transformation of potential vorticity

In the adiabatic case, isentropic surfaces translate in space following air motion, so they are impermeable for air parcels. Under the influence of diabatic heating, weak diffusion of air parcels across the isentropic surfaces begins. It is more convenient to speak of the slow displacement of isentropic surfaces relative to the air, with typical velocities around several millimeters per second. In a reference system linked to isentropic surfaces it is convenient to judge on the diabatic heating impact, observing the relative displacements of equiscalar lines of Ertel's potential vorticity $I = \left(\vec{\omega}_a \cdot \nabla \theta \right) / \rho$ on these surfaces. From the point of view of theory, the starting point is the equation of potential vorticity transformation

$$\frac{D}{Dt} \frac{\vec{\omega}_a \cdot \nabla \theta}{\rho} = \frac{\vec{\omega}_a \cdot \nabla \dot{\theta}}{\rho} \tag{1}$$

where all non-potential forces, including friction, acting on fluid parcels are neglected in the right-hand side. Such an approximation is admissible for free atmosphere conditions, well above the planetary boundary layer, in the framework of which we shall stay.

We write Equation (1) in θ-coordinates, assuming the motion to be quasi-static, and start from the equations of motion on isentropic surfaces in the field of potential forces (see Section 4.1)

$$Du/Dt - lv = -\partial M / \partial x, \quad Dv/Dt + lu = -\partial M / \partial y.$$

By eliminating the M-function from these equations, we arrive at the Friedmann's Equation (9) of Section 1.1, written in θ-coordinates

$$\frac{D}{Dt}\omega_{a\theta} + \omega_{a\theta}\left(\frac{\partial u}{\partial x} + \frac{\partial v}{\partial y}\right) = \frac{\partial \dot{\theta}}{\partial y}\frac{\partial u}{\partial \theta} - \frac{\partial \dot{\theta}}{\partial x}\frac{\partial v}{\partial \theta},$$

where $\omega_{a\theta}$ is the absolute vorticity component orthogonal to an isentropic surface. We write the mass-conservation equation in θ-coordinates in the form

$$\frac{D}{Dt}\rho_\theta + \rho_\theta\left(\frac{\partial u}{\partial x} + \frac{\partial v}{\partial y} + \frac{\partial \dot{\theta}}{\partial \theta}\right) = 0.$$

Eliminating the isentropic velocity divergence $(\partial u/\partial x) + (\partial v/\partial y)$ from the above written equations, we arrive at the equation of transformation of potential vorticity $I = \omega_{a\theta}/\rho_\theta$

$$\rho_\theta\frac{DI}{Dt} = \omega_{a\theta}\frac{\partial \dot{\theta}}{\partial \theta} - \frac{\partial \dot{\theta}}{\partial x}\frac{\partial v}{\partial \theta} + \frac{\partial \dot{\theta}}{\partial y}\frac{\partial u}{\partial \theta}. \tag{2}$$

For large-scale atmospheric processes, this equation gains a simple approximate form (Hoskins *et al.*, 1985)

$$\frac{DI}{Dt} \approx I\frac{\partial \dot{\theta}}{\partial \theta}. \tag{3}$$

Suppose, we apply a localized diabatic heating source, with vertical extent small enough compared to the atmospheric scale-height. On the other hand, this diabatic heating is assumed to be smoothly distributed with altitude in such a way that local Richardson's number remains positive everywhere, and the vertical convection does not start. As a consequence, just above the heat source the air mass enclosed between two neighboring isentropic surfaces increases, and right beneath the heat source it decreases. For a cold source the picture is reversed. By its very definition, the potential vorticity is the specific vorticity charge density per unit mass. Thus, just above the heat source, the potential vorticity values decrease, and anticyclogenesis takes place, and right below the diabatic heating source they increase, i.e., cyclogenesis happens (see Figure 8).

What was said above, could be illustrated using the analysis of Equation (3). Consider a simple problem, one-dimensional in the vertical direction, when the equation in question takes the form

$$\frac{\partial I}{\partial t} + \dot{\theta}\frac{\partial I}{\partial \theta} = I\frac{\partial \dot{\theta}}{\partial \theta}. \tag{4}$$

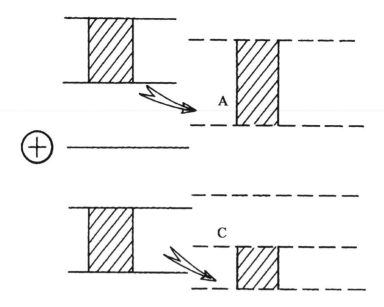

FIGURE 8 Mechanism responsible for anticyclogenesis (A) in a region above a diabatic heating source and for cyclogenesis (C) happening below it, due to diminishing (or, correspondingly, increasing) of vorticity charge concentration within a layer enclosed between two isentropic surfaces. The solid line denotes isentropic surfaces at initial time instant $t = 0$, the dashed line – at a time-moment $t = \tau > 0$. In the case of diabatic cooling, the picture should be reversed.

Let us assume that diabatic heating $\dot{\theta}$ is distributed with height according to the law $\dot{\theta} = f(\zeta)$, where $\zeta = 0$, i.e., ζ is a vertical Lagrangian coordinate, and is switched on at the initial time-instant $t = 0$. This parameterization is based on the assumption that diabatic heating is directly related to some conservative air parcel properties, i.e., to specific concentration of optically active species, including ozone, which absorb the solar radiation. Equation (4) is re-written in the form

$$\frac{\partial}{\partial t}\left(I^{-1}\right) + \frac{\partial}{\partial \theta}\left(\dot{\theta}I^{-1}\right) = 0 \tag{5}$$

and further such a function Z is introduced that $I^{-1} = \partial Z/\partial\theta$ and $\dot{\theta}I^{-1} = -\partial Z/\partial t$. Now, Equation (5) is satisfied identically, and Z becomes a Lagrange invariant

$$\dot{Z} = 0. \tag{6}$$

So, one has that $Z = \varphi(\zeta)$, where φ is a certain scalar function. Therefore, it could be assumed that

$$\dot\theta = f\left[\varphi^{-1}(Z)\right] = F(Z). \tag{7}$$

Equations (6), (7), under the initial condition $\theta = \xi$, $t = 0$ have the solution

$$Z_1(\theta) = Z_0(\xi), \quad \theta = \xi + F\left[Z_0(\xi)\right]t.$$

Thus, one obtains

$$I^{-1} = \frac{dZ_1}{d\theta} = \frac{dZ_0/d\xi}{1 + \left(dF/dZ_0\right)\left(dZ_0/d\xi\right)t}. \tag{8}$$

It is assumed that the initial field of potential vorticity $I_0 = (dZ_0/d\xi)^{-1}$ has the same sign over all the altitudes. For definiteness, let it be $I_0 > 0$. Above the point of diabatic heating maximum $dF/dZ_0 < 0$, and $dF/dZ_0 > 0$ below this point. In the first case, one has the diabatic anticyclogenesis; in the second case, the cyclogenesis, respectively. A 'catastrophe' will happen at the time-instant when the denominator in formula (8) will vanish. After 'switching on' the diabatic heating source, this will occur in a time-interval τ, determined through the condition

$$\tau^{-1} = \left| \frac{dF}{dZ_0} \frac{dZ_0}{d\xi} \right|_{max}.$$

Suppose that at initial time-instant the diabatic heating is distributed over altitude by the Gaussian law

$$\dot\theta\big|_{t=0} = F\left[Z_0(\xi)\right] = A\exp\left\{-\frac{(\xi - \xi_0)^2}{a^2}\right\},$$

and, what is more, the Gaussian curve semi-width a corresponds to $|\xi - \xi_0|$ = 10 K. Now, $\tau = (a/A)\sqrt{e/2}$. Assuming $A = 1$ K·days^{-1}, which suits well the free atmosphere conditions, we arrive at the estimate $\tau \approx 10$ days. Thus, treating time-intervals $\Delta t = 1–2$ days, one can use adiabatic approximation with fair confidence.

Accounting for the horizontal structure of both potential vorticity and diabatic heating rate fields, Equation (3) takes the form

$$\frac{D_\theta}{Dt}\left(I^{-1}\right) = -\frac{\partial}{\partial\theta}\left(\dot\theta I^{-1}\right),$$

where the notation

$$\frac{D_\theta}{Dt} = \frac{\partial}{\partial t} + u\frac{\partial}{\partial x} + v\frac{\partial}{\partial y} \equiv \frac{\partial}{\partial t} + \mathbf{v}\cdot\nabla_\theta$$

for material time-derivative along isentropic surfaces is introduced. Under adiabatic approximation, the derivatives D_θ/Dt and D/Dt become identical. Let us determine the translation velocity $\mathbf{u}$ of the isoline $I = \text{const}$ along isentropic surfaces, starting from the characteristic equation $\partial I/\partial t + \mathbf{u}\cdot\nabla_\theta I = 0$. The diabatically induced propagation velocity $\mathbf{v}_d = \mathbf{u} - \mathbf{v}$ of the isoline $I = \text{const}$, considered relative to the air and taken along isentropic surfaces, is determined through the equation

$$\mathbf{v}_d \cdot \nabla\left(I^{-1}\right) = \frac{\partial}{\partial\theta}\left(\dot\theta I^{-1}\right).$$

Typical $\mathbf{v}_d$ values are of the order of $1\ \text{m·s}^{-1}$. Integrating the resulting equation over altitude, starting from the given isentropic level $\theta = \text{const}$ up to the very top of the atmosphere, where it is assumed that $I^{-1}\theta \to 0$, we shall have

$$I^{-1}\dot\theta = \int_\theta^\infty \left(\mathbf{v}_d \cdot \nabla_\theta \ln I\right) I^{-1} d\theta. \tag{9}$$

Diabatic heating causes a gradual movement of isolines $I = \text{const}$ through the air in the direction of I growing within the entire air column above the chosen equiscalar surface, $\theta = \text{const}$. The integral relation (9) can be put into the ground of a diabatic heating rate $\dot\theta(x, y, \theta, t)$ computation, which is based on available information on dynamic processes at overlying isentropic levels (Agayan et al., 1990; Kurgansky and Tatarskaya, 1990). In practice, using Equation (9) it is possible to restrict oneself by integrating over one–two atmospheric scale-heights. In particular, in the vicinity of the tropopause, the $\dot\theta$-field is determined by dynamic processes in the stratosphere, which opens definite prospects for estimating the stratospheric reverse action on large-scale tropospheric processes (cf. McIntyre and Norton, 1990; Holton *et al.*, 1995).

4.5 General properties of air adiabatic motion

After the principal role of diabatic heating processes in potential vorticity field transformation and the respective conditions of applying the adiabatic approximation to large-scale processes analysis, have been discussed, let us consider some general questions in the theory of adiabatic invariants.

1 Gauge invariance in potential vorticity definition

Along with Ertel's potential vorticity $I = (\vec{\omega}_a \cdot \nabla\theta)/\rho$ there exists an infinite set of adiabatic invariants, linear by the vorticity field

$$\frac{D}{Dt}\frac{\vec{\omega}_a \cdot \nabla f(\theta)}{\rho} = 0, \tag{1}$$

where $f(\theta)$ is a certain differentiable function. This is an immediate consequence of Ertel's (1942) theorem. Every quantity under the material time-derivative sign in Equation (1) could be called the potential vorticity. We impose a sole condition for f to be a monotonic function of its argument. Let $f'(\theta) > 0$, for definiteness. In particular, as it was already stressed in Chapter 1, the choice $f(\theta) = c_p \ln\theta + \text{const}$ is convenient in theoretical studies. Restrict a set of f-functions, assuming that $f(\theta) \to A = \text{const}$ at $\theta \to \infty$, and, what is more, $A = 0$, without the loss of generality. For such a subset of f-functions, the integral

$$\iiint \left\{ (\vec{\omega}_a \cdot \nabla f(\theta))/\rho \right\} \rho \, d\tau,$$

taken over the entire atmospheric volume, remains a finite quantity irrespective of the precise formulation of the boundary conditions at the top of the atmosphere. If the f-function increases monotonously with θ and tends to zero at $\theta \to \infty$, it is negative everywhere. As suggested by Obukhov (1964), we take it in the form $f = \chi(\theta) \equiv -p^*(\theta)/g$, where g is the gravity acceleration and $p^*(\theta)$ is the reference pressure distribution upon isentropic levels, which could be calculated, for example, from the reference atmosphere variables. Namely, Obukhov (1964) introduced the invariant

$$\tilde{\Omega} = (\vec{\omega}_a \cdot \nabla\chi)/\rho.$$

In isentropic coordinates the potential vorticity $\tilde{\Omega}$ is written in the form

$$\tilde{\Omega} = \omega_{a\theta}\left(\rho*_\theta(\theta)/\rho_\theta\right), \quad \rho*_\theta = -g^{-1}dp*(\theta)/d\theta,$$

and in the case when the vertical stratification parameters coincide with those of the reference atmosphere, $\tilde{\Omega}$ is identical to the absolute vorticity $\omega_{a\theta}$.

The existing arbitrariness in the choice of $p*(\theta)$ is removed, for example, if we equate the latter to the average pressure on isentropic surfaces for an actual atmospheric state

$$p*(\theta) = \tilde{p}(\theta) = \iint p d\sigma / \iint d\sigma,$$

where integration is extended over surfaces $\theta = $ const. Pressure $\tilde{p}(\theta)$ is proportional to the atmospheric mass above the given isentropic surface and is a constant of adiabatic motion. The exact concept of Lorenz' available potential energy is formulated with the help of $\tilde{p}(\theta)$ (see Section 4.3).

The second, independent, method of choosing the function $p*(\theta)$ could be based on one consequence from Kelvin's circulation theorem, namely on the coincidence of absolute vorticity fluxes across a given isentropic surface for both actual and reference atmospheric states

$$\iint \omega*_{a\theta} d\sigma = \iint \omega_{a\theta} d\sigma.$$

Assuming that a reference atmospheric state could be achieved with the help of an adiabatic process, with Ertel's potential vorticity conservation in every air parcel, when $\omega*_{a\theta}/\rho*_\theta = \omega_{a\theta}/\rho_\theta$, we write the preceding equality in the form

$$\iint \omega_{a\theta}\left(\rho*_\theta/\rho_\theta\right)d\sigma = \iint \omega_{a\theta} d\sigma$$

or

$$\rho*_\theta = \iint \omega_{a\theta} d\sigma / \iint (\omega_{a\theta}/\rho_\theta) d\sigma.$$

At high and mid latitudes, where $|\omega_{a\theta}| > 0$, this operation is well defined.

These two methods are the most admissible when the problem of reproduction of the field of motion starting from the known potential vorticity field appears. This is the problem of 'inversion' of the potential vorticity field (McIntyre and Norton, 2000). Accounting for the diabatic factors, the explicit time-dependence $p^* = p^*(\theta,t)$ appears in the function p^* found by these methods. Thus, in the course of statistical investigations both these two methods are inconvenient because they lead to a necessity to recompute the function p^* every time. It is more suitable to choose p^* as the climate mean pressure distribution upon isentropic surfaces, as it was proposed in Tatarskaya (1978).

In most papers on the isentropic analysis and diagnosis of large-scale atmospheric processes, including those in the troposphere (e.g., Hartmann, 1995; Magnusdottir and Haynes, 1996), Ertel's potential vorticity I is used directly. This quantity has a good property to enable one to clearly distinguish between air masses of tropospheric and stratospheric origin. Due to a higher extent of stratospheric static stability, the corresponding Ertel's potential vorticity I values are, as a rule, several times as large as in the troposphere. The isoscalar surface $I = I_0$, with $I_0 = (1-2) \times 10^{-6}$ $m^2 \cdot K \cdot s^{-1} \cdot kg^{-1}$ fits with good accuracy the position of the extratropical tropopause (Reed, 1955; Reed and Danielsen, 1959; Danielsen, 1968). It should be emphasized that all morphological peculiarities of the potential vorticity field, among them those which manifest themselves in the potential vorticity gradient field, taken along isentropic surfaces, do exist independently from the choice of the function $f(\theta)$ in (1). In such an invariant manner, with respect to function f, evidence appears, e.g., of the intersection between an isentropic surface and that of the tropopause, where the potential vorticity experiences a jump. However, the question of the appropriate choice of $f(\theta)$-function is more important for the description of the vertical structure of the meteorological field. The traditionally used couple of variables (θ,I) is not very convenient in this respect because Ertel's potential vorticity I increases rapidly with altitude, competing in this respect with potential temperature θ. As the result, the gradients ∇I and $\nabla \theta$ become nearly collinear. Geometrically, the replacement $\theta \rightarrow f(\theta)$ means the change in the slope between the isoscalar potential vorticity surface and the horizontal plane. By choosing an appropriate f-function this angle could be on average made much larger than for equiscalar surfaces with $I = $ const (Figure 9).

2 *Topological invariants*

During adiabatic processes in the atmosphere, the tangent property between isentropic surfaces and isoscalar potential vorticity surfaces is conserved. It is easy to see that under the influence of the replacement $\theta \rightarrow f(\theta)$ in the potential vorticity definition a corresponding tangent point can neither appear nor disappear, if only f is a monotonic function, the latter property

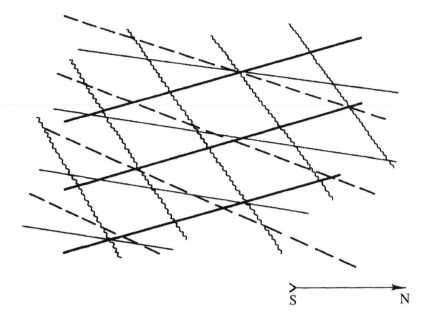

S   N

FIGURE 9 Cross-section in a vertical plane of isentropic surfaces (bold solid lines), equiscalar surfaces of Ertel's potential vorticity defined through potential temperature (thin solid lines), equiscalar surfaces of Ertel's potential vorticity defined via specific entropy (dashed lines), and equiscalar surfaces of potential vorticity, modified following Obukhov (1964) (wavy lines).

being especially compromised. Thus, in future theoretical treatments, the Ertel's potential vorticity I will be used, to be definite. The local structure of isoscalar surfaces $\theta = $ const and $I = $ const is naturally characterized by the vector field $\mathbf{B} = \nabla I \times \nabla \theta$, where $\mathbf{B}$ is a tangent vector to both these surfaces and also to the line of their intersection. Starting from the adiabatic motion equations

$$\frac{\partial \theta}{\partial t} + \mathbf{v} \cdot \nabla \theta = 0, \quad \frac{\partial I}{\partial t} + \mathbf{v} \cdot \nabla I = 0,$$

we compose the local time-derivative of the vector field $\mathbf{B}$:

$$\frac{\partial \mathbf{B}}{\partial t} = \nabla\left(\frac{\partial I}{\partial t}\right) \times \nabla\theta + \nabla I \times \nabla\left(\frac{\partial \theta}{\partial t}\right) = \nabla \times \left(\frac{\partial I}{\partial t}\nabla\theta\right) - \nabla \times \left(\frac{\partial \theta}{\partial t}\nabla I\right)$$

$$= \nabla \times \{\nabla I(\mathbf{v} \cdot \nabla\theta) - \nabla\theta(\mathbf{v} \cdot \nabla I)\} = \nabla \times \{\mathbf{v} \times (\nabla I \times \nabla\theta)\} = \nabla \times \{\mathbf{v} \times \mathbf{B}\}.$$

Because vector **B** is solenoidal and applying a well-known vectorial identity, we arrive at the equation

$$\frac{D\mathbf{B}}{Dt} - (\mathbf{B} \cdot \nabla)\mathbf{v} + \mathbf{B}(\nabla \cdot \mathbf{v}) = 0.$$

Using the Helmholtz' operator helm introduced in Section 1.1, we re-write this equation as helm$\mathbf{B} = 0$, which according to Friedmann (1934) theorem means that **B** is conserved in the adiabatic motion of a compressible fluid, i.e., figuratively speaking, is frozen in a fluid. If at an initial time instant vector **B** vanishes in any material parcel, i.e., a 'tangent event' between the isentropic and isoscalar potential vorticity surfaces happens, this property will survive exactly in the same material parcel and for all subsequent time-instants. The points in which **B** = 0 are called singularities.

It is beneficial to consider I field on isentropic surfaces. In this case, one could speak of isentropic potential vorticity maps (Hoskins *et al.*, 1985). Here, singularities are the points where potential vorticity gradient, taken along an isentropic surface, vanishes. If the singularities are primary (simple), i.e., the corresponding quadratic form constructed from I spatial derivatives is non-degenerative, then they could be of only two types: either centers (extrema of I), or coles (saddles). We restrict ourselves to the consideration of primary singularities exceptionally, because other manifolds, for which the tangency of isoscalar surfaces θ = const and I = const might occur, are structurally unstable. According to a well-known topological assertion, sometimes called the Euler–Poincaré theorem (Arnol'd, 1978), an uneven (odd) number of singularities always exists inside any closed isoline I = const lying in an isentropic surface; what is more, the number of centers exceeds the number of saddles by unity. Under mild diabatic heating and friction, which are permanently present in the atmosphere, the number of singularities might change but only by a multiple of two. Two main types of such elementary reconstructions of the potential vorticity field, when the number of singularities increases by two, i.e., one center and one saddle point are added, are possible. First, that which leads to the low cut off formation (Figure 10b) and, second, that which results in the blocking high cut-off (Figure 10a). As a whole, the frequency of the potential vorticity field topological reconstructions on isentropic surfaces is the measure of non-adiabaticity of atmospheric processes. Changes in the number of singularities are most probable in spring and fall, when the influence of the diabatic factor on the atmospheric general circulation is maximum.

In the case of atmospheric adiabatic motion there are two immediate consequences of the above arguments, to which we would like to draw the reader's attention: (i) the extremum values of potential vorticity on isentropic surfaces, as well as I values in the saddle points, are time-constant,

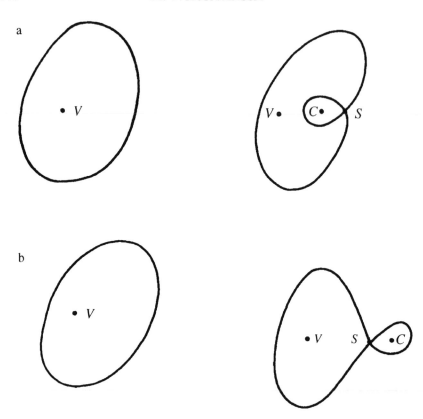

FIGURE 10 Two main types of the reconstruction of Ertel's potential vorticity field *I* on isentropic surfaces: with blocking high cut-off (a) and cyclonic low cut-off (b). Equiscalar lines *I* = const are marked with solid lines; letters *C* and *S* denote the emergence of a new center and a saddle point (cole), respectively; *V* is a singularity which corresponds to the circumpolar vortex.

and (ii) on isentropic surfaces single connected regions, bounded by the closed potential vorticity isolines, are conserved. This could be proved, i.e., *ad absurdum*; actually, violation of the property of a domain to be single connected would result, with necessity, in the appearance of new singularities, the latter being impossible due to the assumption that the motion is strictly adiabatic.

It is worth mentioning that a pioneering paper on numerical weather forecasting (Charney *et al.*, 1950), aiming at the numerical code control and estimation of the degree of applying the model to reality, has suggested to compare the initial and predicted absolute vorticity fields on a 500 hPa surface in order to test the main consequences of absolute vorticity conservation. The latter quantity is the material constant for the barotropic atmospheric model used for numerical forecasting. These consequences are: (i) constancy of the extremum values of vorticity, (ii) conservation of the single connectedness property of domains bounded by absolute vorticity isolines, and, finally, (iii) constancy of area encircled by any closed absolute vorticity isoline. The first two invariants are completely transferred onto the case of a three-dimensional baroclinic atmosphere, as it has been already made above, but the last invariant needs serious modification.

3 Account for compressibility of motion on isentropic surfaces

Strictly speaking, an area enclosed by a contour l = const taken in an isentropic surface is not constant, although sometimes this is implicitly hypothesized and utilized when diagnosing large-scale atmospheric processes (Butchart and Remsberg, 1986). One may speak of approximate area conservation, giving oneself an account for the accuracy of the assumption used. Fortunately, in the three-dimensional baroclinic case there is an invariant, which generalizes the area conservation principle in the two-dimensional problem. According to Kelvin's circulation theorem, the absolute vorticity $\vec{\omega}_a$ flux across a material area Σ taken on an isentropic surface, which is encircled within a reducible closed contour l = const

$$\iint_{\Sigma} \vec{\omega}_a \cdot \mathbf{n}\, d\sigma = \text{invariant},$$

under assumption of the absence of both diabatic heating and non-potential external forces. Here, $\mathbf{n} = \nabla\theta/|\nabla\theta|$ is a unit vector orthogonal to the isentropic surface. Using both absolute vorticity definition $\vec{\omega}_a = \nabla \times \mathbf{v} + 2\vec{\Omega}$ and Stokes' theorem, we re-write this invariant in the form

$$\iint_{\Sigma} 2\vec{\Omega} \cdot \mathbf{n}\, d\sigma + \oint_{L} \mathbf{v} \cdot d\mathbf{l} = \text{invariant}, \tag{2}$$

where a reducible contour L encloses Σ and coincides with an isoline l = const. With the aim to analyze qualitatively a comparative role of the two terms in Equation (2), we assume that contour L is a circle of radius r. If $\bar{v}_l$ is an average value of tangent velocity at L, and $\bar{l}$ is the value of average planetary vorticity $2\vec{\Omega} \cdot \mathbf{n}$ over Σ, we shall have

$$\bar{l}\pi r^2 + 2\pi r \bar{v}_l = \text{invariant}. \tag{3}$$

For small- and mesoscale motions, formally at $r \to 0$, the Earth's rotation is of minor significance and an approximate $2\pi r \bar{v}_l \approx \text{invariant}$ holds. According to it, the tangent velocity changes according to angular momentum conservation law demands, at the expense of horizontal compressibility, i.e., of the changes in radius r. For large-scale motion, with $r \sim 1000$ km, both terms in Equation (3) are equally important, though at a first glance it might seem that the first term greatly exceeds the second term in magnitude. The matter is that such large-scale motion is nearly incompressible in isentropic surfaces due to the closeness of the meteorological fields to the geostrophic balance (cf. Chapter 2). Thus, an additional $\pi r^2 \approx \text{invariant}$ exists which, taken with an appropriate coefficient, could be subtracted from Equation (3) in order to arrive at

$$\left(\bar{l} - \bar{l}_0\right)\pi r^2 + 2\pi r \bar{v}_l \approx \text{invariant}. \tag{4}$$

Here, $\bar{l}_0$ is a reference Coriolis parameter. For large-scale motions $\left(\bar{l} - \bar{l}_0\right)/\bar{l}_0 = O(10^{-1})$ and both terms in Equation (4) are of the same order of magnitude. The motions of the gravest, or planetary, scale, with $r \geq 3000$ km, could also be of interest. Here, the first term in Equation (3) becomes dominating, so $\bar{l}\pi r^2 \approx \text{invariant}$, and the horizontal compressibility starts to play a decisive role again. The given situation could be characterized as the limiting case of strong atmospheric baroclinicity. These dominating baroclinic factors determine the vertical wind shear and in such a way impose the upper limit on possible horizontal wind shears. In other words, they suppress the 'fluid dynamic induction' mechanism, which is responsible for the generation of the relative vorticity at the expense of the planetary one. Here, we arrive at an approximation

$$\iint_\Sigma 2\vec{\Omega} \cdot \mathbf{n} d\sigma \approx \text{invariant}, \tag{5}$$

and by the theorem on the mean values $\bar{l}\Sigma \approx \text{invariant}$, where $\bar{l}$ is the average value of $2\vec{\Omega} \cdot \mathbf{n}$ over Σ. As a consequence, along with its poleward excursion, the area of a domain encircled by the closed contour $l = \text{const}$ decreases, and in the course of a equatorward excursion it increases. It is of interest to consider an example of a ring domain, having the isolines $l = l_1, l_2$ as the lateral boundaries, which encircle the pole and are oriented nearly along the latitudinal circles. If the outer diameter of the ring is diminished, its area will also decrease, but due to the meridian convergence the ring's width will increase, though not so strongly, as if it were for incompressible motion. To gain a qualitative impression of this effect, let us consider the case of strictly zonal circulation, when the isolines $I = I_1$, I_2 coincide with the latitudinal circles $\varphi = \varphi_1$, φ_2, and write down the condition for conservation of the approximate invariant (5) for such a ring

$$\iint_\Sigma 2\vec{\Omega} \cdot \mathbf{n} d\sigma = 2\pi a^2 \int_{\varphi_1}^{\varphi_2} 2\Omega \sin\varphi \cos\varphi d\varphi = 2\pi\Omega a^2 \left(\sin^2 \varphi_2 - \sin^2 \varphi_1\right)$$

$$= 2\pi\Omega a^2 \sin(\varphi_2 - \varphi_1)\sin(\varphi_2 + \varphi_1) \approx \text{invariant.}$$

If the ring is thin enough, i.e., $\Delta\varphi = \varphi_2 - \varphi_1 \ll \pi/2$, then $\Delta\varphi \sin(\varphi_2 + \varphi_1)$ $\approx$ invariant, and its width is minimum at $(\varphi_2 + \varphi_1)/2 = 45°$. Along with any changes in the outer diameter of the ring, leading to either its increase or decrease, the ring becomes thicker in any case. If we start from the principle of such a ring's area conservation, we would have

$$\iint_\Sigma d\sigma = 2\pi a^2 \int_{\varphi_1}^{\varphi_2} \cos\varphi d\varphi = 2\pi a^2 \left(\sin \varphi_2 - \sin \varphi_1\right)$$

$$= 4\pi a^2 \sin\frac{\varphi_2 - \varphi_1}{2} \cos\frac{\varphi_2 + \varphi_1}{2} = \text{invariant.}$$

Likewise, for a thin ring $\Delta\varphi \cos((\varphi_2 + \varphi_1)/2) \approx$ invariant and it has the minimum width at low latitudes, and becomes monotonously thicker when its outer diameter diminishes. Both the principle of area conservation and the principle of planetary vorticity flux conservation (5) are two different constructive approximations to the precise principle of absolute vorticity flux conservation (2). As we have just seen for a simple example of the ring domain, these principles lead to different consequences. Which principle is more advantageous for the diagnostics of large-scale atmospheric processes could be said only staying on the grounds of further isentropic analysis.

4   Invariant flux tubes and their deformation

A family of isentropic surfaces $\theta = $ const, intersecting with isoscalar Ertel's potential vorticity surfaces $I = $ const, forms a set of invariant tubes, or solenoids. During adiabatic processes, occurring in the potential force field, these tubes behave as material objects, i.e., they are frozen in a fluid and move following fluid parcels. They can be either closed or end on fluid boundaries. In this respect they are quite analogous to common vorticity tubes in barotropic fluid dynamics. Infinitely thin (θ,l)-tubes are completely characterized by the vector $\mathbf{B} = \nabla l \times \nabla \theta$, which is conserved in the processes in question, i.e., obeys the governing equation

$$\partial \mathbf{B}/\partial t = \nabla \times (\mathbf{v} \times \mathbf{B}),$$

or

$$\frac{D\mathbf{B}}{Dt} = (\mathbf{B} \cdot \nabla)\mathbf{v} - \mathbf{B}(\nabla \cdot \mathbf{v}). \tag{6}$$

As it is well-known (Pedlosky, 1987), Equation (6) accounts for two main effects: (i) the lengthwise stretching, and (ii) the transversal tilting of such a tube. Indeed, let us orient the x-axis of Cartesian (x,y,z) coordinates, with unit orthogonal vectors $(\mathbf{i},\mathbf{j},\mathbf{k})$, along the vector $\mathbf{B}$, thus assuming that $\mathbf{B} = B\mathbf{i}$. The right-hand side of equation (6) is written in the form

$$B\frac{\partial}{\partial x}\left(u\mathbf{i} + v\mathbf{j} + w\mathbf{k}\right) - B\mathbf{i}\left(\frac{\partial u}{\partial x} + \frac{\partial v}{\partial y} + \frac{\partial w}{\partial z}\right)$$

$$= B\frac{\partial v}{\partial x}\mathbf{j} + B\frac{\partial w}{\partial x}\mathbf{k} - B\mathbf{i}\left(\frac{\partial v}{\partial y} + \frac{\partial w}{\partial z}\right)$$

The first two terms describe the tilting of a tube and the last term, its stretching. A screwing effect remains not taken into account. Such a problem does not appear in the framework of basic fluid dynamics because of fluid homogeneity. From the very beginning, we make a reservation that stable vertical atmospheric stratification imposes a severe restriction on such screwing. The latter is allowed only in the form of shearing, along isentropic surfaces, of the parallelogram, which according to a well-known theorem on infinitesimal quantities could always approximate the cross-section of an infinitely thin (θ,l)-solenoid. A good illustration of this is the shift of a pack of cards, laying on a table, under an applied effort of fingers.

A strict theory of (θ,l)-solenoid screwing is rather complicated. If we use the Cartesian coordinates x_i ($i = 1, 2, 3$), then such a screwing is described by a quantity

$$\frac{\partial v_i}{\partial x_k} \frac{\partial I}{\partial x_i} \frac{\partial \theta}{\partial x_k},$$

where the repeated indices denote summation. Extracting the symmetric (with respect to the commutation of indices i and k) and antisymmetric parts, one gets

$$\frac{1}{2}\left(\frac{\partial v_i}{\partial x_k} + \frac{\partial v_k}{\partial x_i}\right)\frac{\partial I}{\partial x_i} \frac{\partial \theta}{\partial x_k} + \frac{1}{2}\left(\frac{\partial v_i}{\partial x_k} - \frac{\partial v_k}{\partial x_i}\right)\frac{\partial I}{\partial x_i} \frac{\partial \theta}{\partial x_k}.$$

Using an adiabatic assumption, the symmetric part of this expression is presented in the form $(1/2)D(\nabla I \cdot \nabla \theta)/Dt$ and the antisymmetric part as $(1/2)(\mathbf{B} \cdot \nabla \times \mathbf{v})$, respectively. For large-scale atmospheric processes, it could be shown that $D(\nabla I \cdot \nabla \theta)/Dt \approx \mathbf{B} \cdot \nabla \times \mathbf{v}$, with an accuracy of about 1%. Thus, the invariant tube screwing effect could be equally measured in terms of a material time-rate of the scalar product of the vectors ∇I and $\nabla \theta$, and by the quantity $\mathbf{B} \cdot \nabla \times \mathbf{v}$.

A simplified treatment of (θ, I)-solenoid screwing could be proposed in isentropic θ-co-ordinates, where the equality

$$\frac{D}{Dt}\frac{\partial I}{\partial \theta} = -\frac{\partial u}{\partial \theta}\frac{\partial I}{\partial x} - \frac{\partial v}{\partial \theta}\frac{\partial I}{\partial y}$$

holds for adiabatic processes. This is an immediate consequence of the potential vorticity conservation law. Using the thermal wind relations $\partial u/\partial \theta = -l^{-1}(\partial \Pi/\partial y)$, $\partial v/\partial \theta = l^{-1}(\partial \Pi/\partial x)$, we shall have

$$\frac{D}{Dt}\frac{\partial I}{\partial \theta} = -l^{-1}J(I,\Pi),$$

where J denotes the Jacobian operator. The latter vanishes when and only when $\Pi = \Phi(I,\theta)$, with Φ as an arbitrary function, i.e., when three gradients ∇I, $\nabla \theta$ and $\nabla \Pi$ are complanar and the corresponding mixed product vanishes. The effect of (θ, I)-solenoid screwing is determined by the mutual position of the invariant solenoids and the thermodynamic (Π, θ)-solenoids which, according to the Bjerknes' circulation theorem, determine the vorticity production rate and are permanently present in the atmosphere (Figure 11). Fortunately, for the invariant tubes of finite cross-section, which are constructed based on the observational data, the screwing effect is of secondary significance because of the flatness of these tubes in vertical direction,

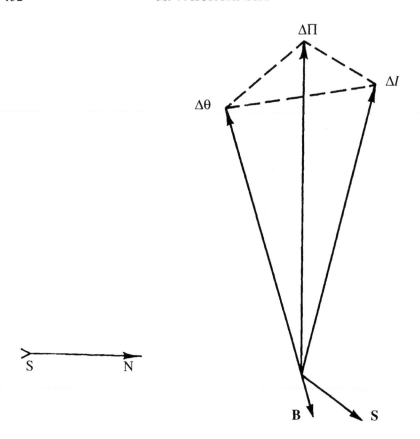

FIGURE 11   Spatial orientation of the main vectors determining invariant tubes and their screwing. In a general case the vectors $\nabla\theta$, ∇I and $\nabla\Pi$ are non-complanar; $\mathbf{S} = \nabla\Pi\times\nabla\theta$ is the baroclinicity vector; $\mathbf{B} = \nabla I\times\nabla\theta$. The screwing effect is absent when and only when the vectors $\nabla\theta$, ∇I, and $\nabla\Pi$ become complanar.

if only they are far away from degeneration. In other words, the angle λ between the isoscalar surfaces $\theta = $ const and $I = $ const should be not small.

Under the influence of diabatic heating and non-potential forces the characteristic vector $\mathbf{B}$ transforms according to the equation[1]

$$\text{helm}\mathbf{B} = \nabla\times\left(\dot{I}\,\nabla\theta - \dot{\theta}\nabla I\right).$$

Due to the Friedmann's theorem, in order for the invariant tubes to be conserved, the adiabatic condition plus the existence of external force potential $\dot{\theta} = \dot{I} = 0$ should be assumed, or a less restrictive condition

$$\dot{I}\,\nabla\theta - \dot{\theta}\,\nabla I = \nabla\Psi,$$

could be adopted which includes an assumption of non-adiabaticity of a special type. Here, Ψ is an arbitrary function of both spatial coordinates and time. It is clear that in reality $\Psi = \Psi(\theta, I, t)$ and, which is more,

$$\dot{I} = \partial\Psi/\partial\theta, \quad \dot{\theta} = -\partial\Psi/\partial I. \tag{7}$$

Equations (7) formally coincide with the Hamiltonian equations in classical mechanics. For steady air motion, Ψ is a Lagrangian invariant. The above arguments allow an analogy with the general theory for Clebsch transformation in basic fluid dynamics (Lamb, 1945).

In an adiabatic case, passing from Ertel's potential vorticity $I = (\vec{\omega}_a \cdot \nabla\theta)/\rho$ to the modified potential vorticity $\tilde{I} = (\vec{\omega}_a \cdot \nabla f(\theta))/\rho$ the slope of isoscalar potential vorticity surfaces to the horizontal plane changes. All adiabatic motion properties remain unchanged. Under general circumstances, when non-adiabatic factors play the role, this is not the case. In particular, let us clarify how it is necessary to transform the vertical coordinate θ in order to arrive again at the condition of non-adiabaticity of a special type

$$\dot{\tilde{I}} = \partial\tilde{\Psi}/\partial\varphi, \quad \dot{\varphi} = -\partial\tilde{\Psi}/\partial\tilde{I}$$

in new coordinates $(\varphi(\theta), \tilde{I} = f'(\theta)I)$, where $\tilde{\Psi} = \tilde{\Psi}(\varphi, \tilde{I}, t)$ is a certain new function. The corresponding canonical transformation (Leech, 1958) is performed with the help of a generating function $F(\theta, \tilde{I}, t)$, which satisfies

---

[1] To derive this equation, one has to start from the general commutation formula (Ertel, 1960; Hollmann, 1964)

$$\frac{D}{Dt}[\rho^{-1}\nabla\Theta_1 \cdot (\nabla\Theta_2 \times \nabla\Theta_3)] = \rho^{-1}\nabla\dot{\Theta}_1 \cdot (\nabla\Theta_2 \times \nabla\Theta_3)$$

$$+ \rho^{-1}\nabla\Theta_1 \cdot (\nabla\dot{\Theta}_2 \times \nabla\Theta_3) + \rho^{-1}\nabla\Theta_1 \cdot (\nabla\Theta_2 \times \nabla\dot{\Theta}_3),$$

valid for three arbitrary scalar fields Θ_1, Θ_2, Θ_3 (ρ is the compressible fluid density). The further steps are: (i) to substitute $\Theta_2 = I$, $\Theta_3 = \theta$; (ii) to use successively $\Theta_1 = x, y, z$ (which are the Cartesian coordinates) and (iii) to combine the three resulting scalar equations into a vectorial one (see Kurgansky and Tatarskaya, 1987; Kurgansky and Pisnichenko, 2000).

the equation $DF/Dt = I\dot{\theta} - \Psi + \varphi\dot{\tilde{I}} + \tilde{\Psi}$. Here, $I = \partial F/\partial\theta$, $\varphi = \partial F/\partial\tilde{I}$, $\tilde{\Psi} - \Psi = \partial F/\partial t$. Assuming that $F = \Phi(\theta)\tilde{I}$, we obtain $I = \Phi'\tilde{I}$, $\varphi = \Phi$. So, it is necessary to put $(\varphi')^{-1} = f'$. In this case the Jacobian operator of the transformation from the variables (θ, I) to the variables $(\varphi, \tilde{I})$ is equal to unity.

4.6   Isentropic potential vorticity maps and invariant flux tubes

Let us illustrate the theoretical constructions of the preceding Sections by examples of potential vorticity field calculations on isentropic surfaces based on observational data. Hereafter, the adiabatic invariant $\tilde{\Omega}$ defined after Obukhov (1964)

$$\tilde{\Omega} = \rho^{-1}\vec{\omega}_a \cdot \nabla\left(- p*(\theta)/g\right)$$

is considered as potential vorticity. In isentropic coordinates, it takes the form

$$\tilde{\Omega} = \omega_{a\theta}\, \rho*_\theta/\rho_\theta, \quad \rho*_\theta = -g^{-1}\, dp*(\theta)/d\theta.$$

The data archive of the First Global 1978–79 GARP Experiment (FGGE), the so called level IIIa (NMC[2] data), has been used. Here, GARP is the abbreviation for the Global Atmospheric Research Programme. Data on the geopotential, temperature and wind speed were presented for the Northern and Southern hemispheres at 12 standard isobaric levels (1000, 850, 700, 500, 400, 300, 250, 200, 150, 100, 70 and 50 hPa) in the grid points, with a 2.5° step in latitude and longitude. First, the potential vorticity field $\tilde{\Omega}$ was calculated on isobaric surfaces with the help of an approximate formula (1) from Section 4.2, written in spherical coordinates, and then it was interpolated onto isentropic surfaces. To compute the atmospheric static stability and to interpolate the variables in the vertical direction, cubic splines were used. The values of pressure p for the reference atmospheric model, averaged over all seasons, were used to construct the $p*(\theta)$ function. This method is described in more detail in Kurgansky and Tatarskaya (1987, 1990) and Obukhov et al. (1988).

In the papers just cited, large-scale synoptic processes in the troposphere were analyzed. Primarily, charts of potential vorticity $\tilde{\Omega}$ on the so called

---

[2] National Meteorological Center, USA; now the National Centers for Environmental Prediction (NCEP).

main isentropic surfaces were drawn. These are the isentropic surfaces, with $\theta = 300-315$ K, which could be followed inside the troposphere over the whole latitudinal range, from the equator to the poles, not quitting it either to the stratosphere above or 'underground' below. In Figure 12, as an example, the potential vorticity distribution on the isentropic surface $\theta = 307$ K in the Northern Hemisphere on December 24, 1978 at 0 h GMT is shown. For comparison, the geopotential height of a 500 hPa level for the same date is also presented (Kurgansky and Tatarskaya, 1987). Comparison of the maps clearly shows that a blocking formation (a vortex pair constituted from an anticyclone and cyclone) in Figure 12a over the Northern Atlantic, indicated by arrow I, corresponds to a similar (by configuration) structure in Figure 12b (near arrow I). The latter has low potential vorticity values, less than 0.5×10^{-4} s^{-1}, in the northern part where the anticyclone is situated, and high potential vorticity, up to 3×10^{-4} s^{-1}, in the southern cyclonic part. In the course of blocking formation the meridional air mass exchange occurs: an arrow in Figure 12b clearly shows how subtropical air penetrates into polar regions. A strong cyclone over Northern Siberia (arrow II in Figure 12a) corresponds to the region with the highest potential vorticity values of 8×10^{-4} s^{-1} for the date in question. Figure 13 presents the distribution of potential vorticity on three various isentropic surfaces in the Southern Hemisphere, and Figure 14 gives the meridional cross-section of potential temperature and potential vorticity fields, through both the Northern and Southern Hemisphere. Isentropic potential vorticity maps are well-adequate to baric topography charts, but are more informative in the respect that every such a map, due to static stability contribution to the values of $\tilde{\Omega}$, reflects the linkage of meteorological processes between different baric levels. Isentropic potential vorticity maps have good spatial resolution, they are relief, allow one to distinguish clearly the impact of the horizontal advection of air masses on the formation of weather peculiarities. Because the potential vorticity behaves like a quasi-Lagrangian material constant, these maps might be an ideal tool for a classical meteorologist-synopticist, who traditionally thinks in terms of individual air mass transformations. Currently, it should be stated that only lone and humble attempts are made to adapt these maps to the practice of operational weather analysis and forecasting. These maps find much broader applications in studies of minor gaseous constituents in the stratosphere, where an analogy between quasi-Lagrangian behavior of the potential vorticity, on the one hand, and specific concentration of chemical species, on the other hand, is used effectively.

Coordinate surfaces $\theta =$ const and $\tilde{\Omega} =$ const divide the atmosphere into invariant flux tubes, along which air parcels travel during adiabatic processes. From an infinite set of such tubes the so called reference invariant flux tube has been chosen (Obukhov et $al.$, 1988). This tube is generated by the intersection of the isoscalar surfaces $\theta = 305, 310$ K and $|\tilde{\Omega}| = 1\times10^{-4}$,

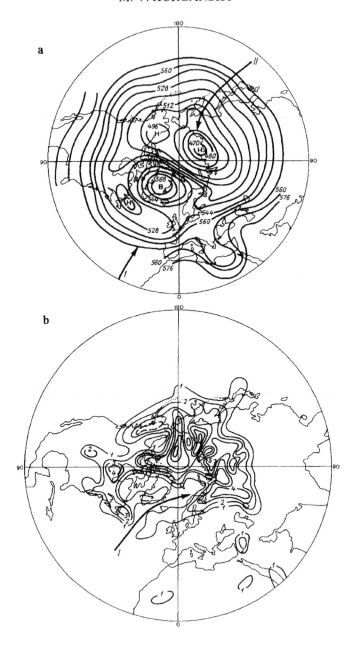

FIGURE 12 The field of the geopotential height (in decameters) on a 500 hPa surface (a), and the distribution of modified potential vorticity (in units of $10^{-4}\,\mathrm{s}^{-1}$) on a 307 K isentropic surface (b) over the Northern Hemisphere on December 24, 1978. The maps are bounded with the latitudinal circles 20 N (a) and 16 N (b); L and H are the regions of low and high pressure, respectively. The meaning of arrows I and II is explained in the text. (From Kurgansky and Tatarskaya, 1987.)

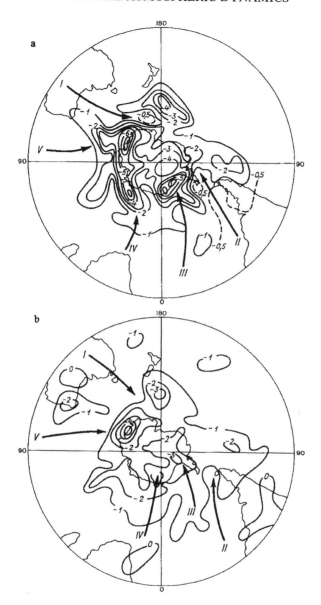

FIGURE 13 Modified potential vorticity distribution on 307 K (a), 327 K (b) and 347 K (c) isentropic surfaces over the Southern Hemisphere on June 2, 1979 (in units of -10^{-4} s^{-1}, the potential vorticity being of the opposite sign to that in the Northern Hemisphere). The maps are bounded with the latitudinal circle 20 S. Arrows *I* and *II* show the regions with low potential vorticity (by their absolute value), which correspond to anticyclones; arrows *III*, *IV* and *V* denote high potential vorticity domains (cyclones). The charts clearly demonstrate how well the potential vorticity field correlates at different altitudes. (From Kurgansky and Tatarskaya, 1987.)

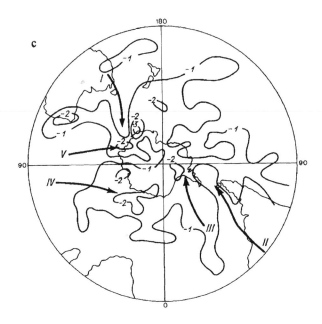

FIGURE 13 (c)

2×10^{-4} s^{-1} ($\tilde{\Omega} > 0$ in the Northern Hemisphere and $\tilde{\Omega} < 0$ in the Southern Hemisphere). In distinction to all other invariant flux tubes this reference tube may be followed over the entire hemisphere, without fragmentation onto separate regional structures, and reflects air circulation within the middle troposphere over high and middle latitudes during adiabatic processes. Besides, it is situated just inside the zone of maximum values of the modulus of potential vorticity gradient, being taken along isentropic surfaces. Thus, this reference tube has a minimum width among all other tubes, with the same difference $\Delta\tilde{\Omega}$ across the tube, which enables one to monitor the meridional air displacements with sufficient resolution.

Strictly speaking, one can consider an invariant flux tube as a material object only when the air circulation time about the closed tube does not exceed, at least, the tube's transformation time due to diabatic heating. An estimate shows that for the reference invariant tube this demand is not satisfied, generally speaking, and this tube consists of different air parcels in different time instants, being a system open by mass. Thus, a mixed Eulerian–Lagrangian approach (cf. Pierrehumbert, 1991) seems to be the most appropriate approach to describe its temporal evolution.

A convenient form of presenting such a three-dimensional object is its projection on the Earth's surface map. Figure 15 shows a reference tube reflection on the Northern hemisphere map for the period December 25, 1978 to January 8, 1979 (Obukhov *et al.*, 1988). Arrow *1* shows a diffluent type of blocking over the Northern Atlantic, and arrow *2* that of meridional

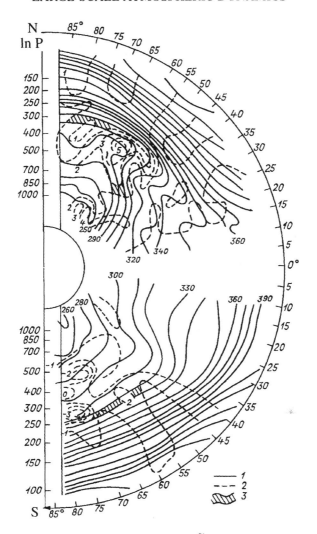

FIGURE 14   Modified potential vorticity $\tilde{\Omega}$ and potential temperature θ distribution in the $\lambda = 60$ W meridional plane on December 24, 1978; isolines $\theta = $ const (1), isolines $\tilde{\Omega} = $ const (2), "reference invariant tube" cross-section (3). (From Kurgansky and Tatarskaya, 1990a.)

type over Rocky Mountains and Pacific Ocean; arrow *3* denotes the sector where the cyclonicity is established. The genesis of cut-off cyclonic lows could be observed (Figure 15e, f), as well as the breakdown of the tube into two fragments (Figure 15d), the latter showing the extreme extent for the topological reconstruction of the potential vorticity field of this particular type.

As another example taken from Kurgansky and Tatarskaya (1990), Figure 16 depicts a reference invariant tube reflection onto the Northern

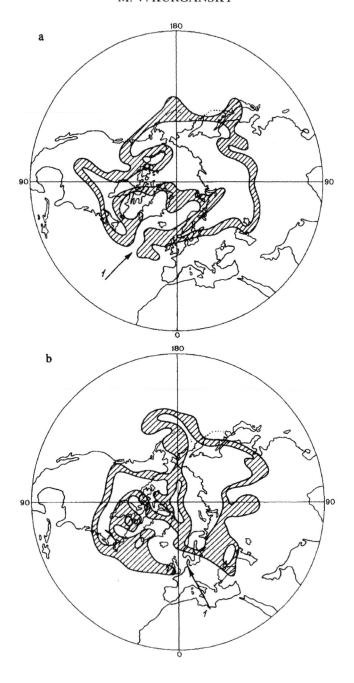

FIGURE 15 The temporal evolution of the 'reference invariant tube': December 24, 1978 (a), December 28, 1978 (b). (From Obukhov *et al.*, 1988.)

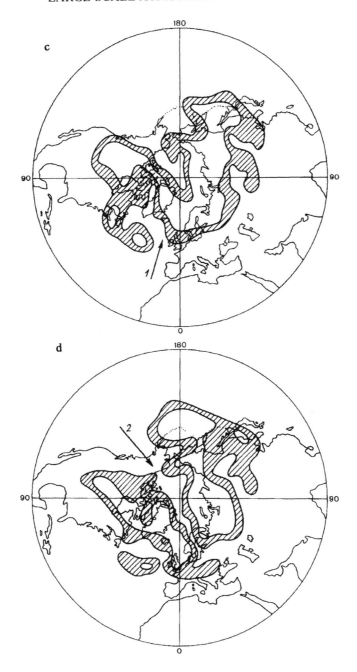

FIGURE 15 The temporal evolution of the 'reference invariant tube':
December 30, 1978 (c), January 1, 1979 (d). (From Obukhov *et al.*, 1988.)

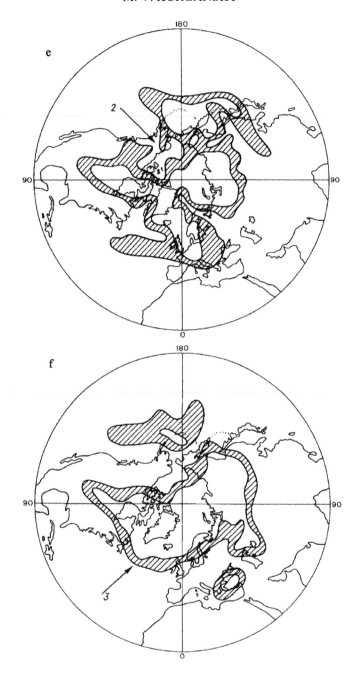

FIGURE 15 The temporal evolution of the 'reference invariant tube': January 2, 1979 (e), January 8, 1979 (f). (From Obukhov *et al.*, 1988.)

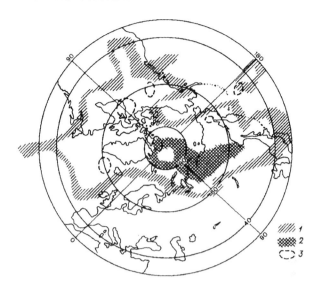

FIGURE 16 A reflection of the 'reference invariant tube' onto the Northern Hemisphere map on February 22, 1979, 3 h GMT; 'reference invariant tube' (1), 'invariant tube', generated by isoscalar surfaces $\tilde{\Omega} = 4 \times 10^{-4}$ s^{-1} and $\tilde{\Omega} = 5 \times 10^{-4}$ s^{-1} (2), cold centers on the 1000/500 hPa relative topography map (3). (From Kurgansky and Tatarskaya, 1990a.)

Hemisphere map on February 22, 1979. At the end of February 1979, a major stratospheric warming took place, which is essentially the strato-spheric circumpolar vortex destruction, being accompanied with potential vorticity meridional mixing along isentropic surfaces (McIntyre and Palmer, 1983, 1984; Hess, 1991). In the troposphere, this phenomenon was accompanied by well-developed cyclonic processes over Eurasia and North America. On relative topography maps 1000/500 hPa, describing the mean temperature of an air column, bounded from above and below by isobaric surfaces 500 and 1000 hPa, respectively, certain strong cold centers were observed in these regions. Because zonal processes (as a direct consequence of well-expressed high-latitude cyclonicity) were well-developed over the entire hemisphere in that period, the reference invariant tube reflection onto the hemispheric map has a latitudinal orientation, as is seen in Figure 16.

As it has been stressed above, blocking formations regularly develop in the atmosphere, which distort the zonal flow, and transform it into the meridional jet stream over certain hemispheric sectors. When compared with routine baric topography maps, the reference invariant tube reflects very clearly the development of this process. Figure 17 shows the behavior of the reference invariant tube during the period of Indian summer. During the Indian summer of 1979, blocking anticyclones were persistent for one

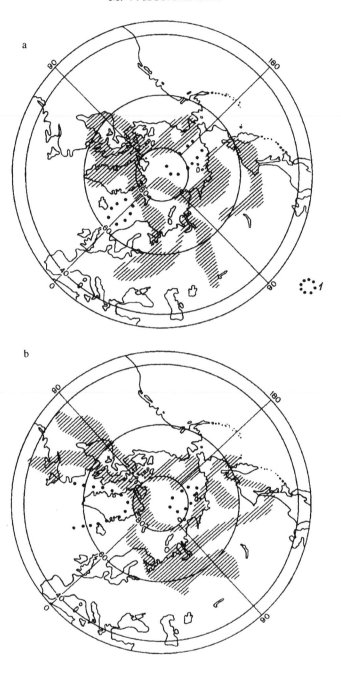

FIGURE 17 A reflection of the 'reference invariant tube' onto the Northern hemisphere map during the autumn of 1979: October 6 (a), October 7 (b). (From Kurgansky and Tatarskaya, 1990.)

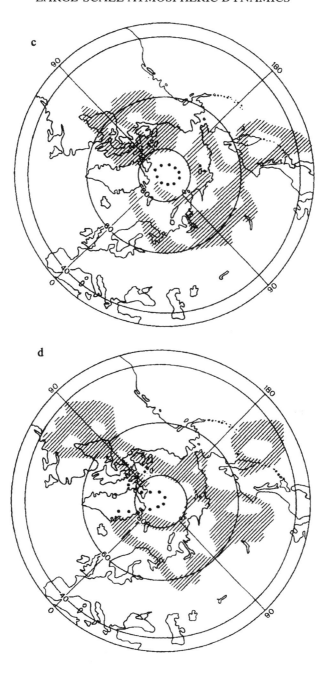

FIGURE 17 A reflection of the 'reference invariant tube' onto the Northern hemisphere map during the autumn of 1979: October 8 (c), October 9 (d). (From Kurgansky and Tatarskaya, 1990.)

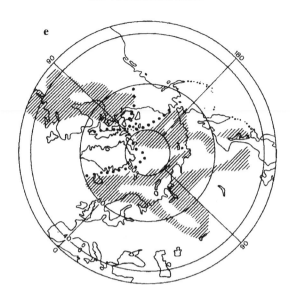

FIGURE 17 A reflection of the a 'reference invariant tube' onto the Northern hemisphere map during the autumn of 1979: October 10 (e). (From Kurgansky and Tatarskaya, 1990.)

month and a half over Eurasia and North America and they were so strong that for certain dates the warm centers on relative topography 1000/500 hPa maps (dotted lines) were noticed not only over continents but over the Northern Pole, where air masses blow over the pole directly. Here, the reference invariant tube changes its topological structure, breaking into two fragments. These topological reconstructions are reliable, because the difference of the potential vorticity values taken in characteristic field points (singularities) exceeds the error of its computation ($\approx 0.2 \times 10^{-4}$ s^{-1}) several times. Transitions of the reference invariant tube from one topological type to another happened repeatedly during the period in question. It could be explained by the choice of the final period of Indian summer, when along with the strengthening cyclonic processes the circumpolar cold center was also amplified, and a unified reference tube was temporarily re-established, though did not gain its classical form, like that in Figure 16. The reference invariant tube breaks probably correlate with the moments of maximum individual temporal changes of adiabatic invariants, when diabatic factors have the maximum influence on general atmospheric circulation.

Thus, temporary analysis of the evolution of the reference invariant tube enables us to monitor more clearly the large-scale atmospheric dynamics and to estimate qualitatively the degree of importance of diabatic factors. An essential problem remains to estimate the impact of these factors quantitatively.

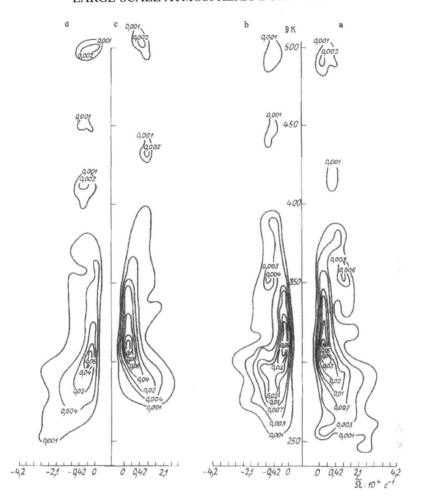

FIGURE 18 Air mass distribution on potential temperature and modified potential vorticity normalized to unity; Northern Hemisphere, December 24, 1979 (a), Southern Hemisphere, December 24, 1979 (b), Northern Hemisphere, June 2, 1979 (c), Southern Hemisphere, June 2, 1979 (d). (From Kurgansky and Tatarskaya, 1987.) Note that figures are numbered from right to left.

4.7 Distribution of adiabatic invariants in the atmosphere

Air mass distribution $\mu(\theta,\tilde{\Omega})$ on θ and $\tilde{\Omega}$ values is a constant of adiabatic atmospheric motion. The quantity $dm = \mu(\theta,\tilde{\Omega})d\theta d\tilde{\Omega}$ is a relative portion of the total atmospheric mass, enclosed within an infinitely thin tube formed by the intersection of the isoscalar surfaces $\theta,\theta + d\theta = $ const and $\tilde{\Omega},\tilde{\Omega} + d\tilde{\Omega}$ = const. It could be statistically treated as a probability to discover an air

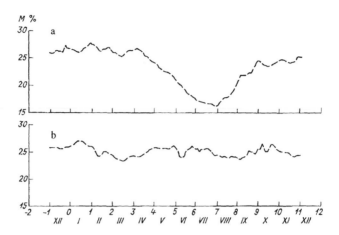

FIGURE 19 Temporal evolution for the entire FGGE period of the northern hemispheric air mass portion M, which falls into the modified potential vorticity range $0.75 \times 10^{-4} \text{ s}^{-1} < \tilde{\Omega} < 2 \times 10^{-4} \text{ s}^{-1}$ (a), and the corresponding part M of the southern hemispheric air mass for $-2 \times 10^{-4} \text{ s}^{-1} < \tilde{\Omega} < -0.75 \times 10^{-4} \text{ s}^{-1}$ (b). The dates starting from December 1, 1978 up to November 30, 1979 are plotted on the horizontal axis. (From Kurgansky and Tatarskaya, 1990.)

mass with θ and $\tilde{\Omega}$ values falling into a given range of variations of their magnitudes (Obukhov, 1964). In equal terms, $\mu(\theta, \tilde{\Omega})$ is the probability density of θ and $\tilde{\Omega}$ values for a randomly chosen air parcel. The pioneering computations of the μ-function, based on limited empirical material, have been performed by A. Karunin at the Institute of Atmospheric Physics in Moscow (see Monin, 1972). Further computations made in Kurgansky and Tatarskaya (1987, 1990), provided information on the global distribution of θ and $\tilde{\Omega}$ in the atmosphere for the entire FGGE period. Examples of the μ-distribution are presented in Figure 18. Seasonal changes are clearly seen and, moreover, the summer μ-distribution for the Northern Hemisphere is the 'sharpest', showing a tendency towards air mass concentration in the limited range of potential vorticity values.

Aiming to reduce errors, induced both by the hidden influence of diabatic heating and/or friction and computational imperfections, and to present information in a more compressed form, a temporal evolution of the Northern and Southern hemispheric air mass portion M, separately, falling into the range $0.75 \times 10^{-4} \text{ s}^{-1} < |\tilde{\Omega}| < 2 \times 10^{-4} \text{ s}^{-1}$ for all θ values, is given in Figure 19 (Kurgansky and Tatarskaya, 1990). A running 7-day averaging has been applied to smooth the curves. A clear asymmetry between the Northern and Southern Hemisphere is evident. A well-pronounced seasonal course in M-values is observed only in the Northern Hemisphere and is practically absent in the Southern Hemisphere.

Table 1 Monthly-mean (FGGE period) informational entropy values for the atmospheres of the Northern Hemisphere H_N and of the Southern Hemisphere H_S.

Year, month		H_N	H_S	Year, month		H_N	H_S
1978	XII	0.745	0.706	1979	VI	0.622	0.754
1979	I	0.754	0.708		VII	0.623	0.746
	II	0.752	0.715		VIII	0.629	0.745
	III	0.747	0.722		IX	0.640	0.744
	IV	0.723	0.722		X	0.657	0.722
	V	0.664	0.743		–	–	–

A comprehensive cumulative characteristic of $\mu(\theta,\tilde{\Omega})$ distribution is its informational entropy defined after Shannon (Tatarskaya, 1978; Kurgansky and Tatarskaya, 1987)

$$H = -K \iint \mu\left(\theta,\tilde{\Omega}\right) \ln \mu\left(\theta,\tilde{\Omega}\right) d\theta d\tilde{\Omega}, \tag{1}$$

where $K > 0$ is a normalization factor. A non-essential additive constant has been omitted. The actual μ-distribution is calculated in the form of a two-dimensional table, with the total cell number equal to N. If the condition $\sum_{\theta}\sum_{\tilde{\Omega}} \mu(\theta,\tilde{\Omega}) = 1$ is satisfied, where all the cells are summed up, then the informational entropy is determined by the formula

$$H = -(\ln N)^{-1} \sum_{\theta}\sum_{\tilde{\Omega}} \mu\left(\theta,\tilde{\Omega}\right) \ln \mu\left(\theta,\tilde{\Omega}\right). \tag{2}$$

Entropy H defined in this way reaches its maximum value, which is equal to unity, in the case of uniform air mass distribution between N cells. In Kurgansky and Tatarskaya (1987) the H-values have been computed with the use of formula (2) for the Northern and Southern Hemisphere, separately. They are denoted as H_N and H_S, correspondingly. The monthly-mean values for 11 months of the FGGE period are given in Table 1. In the winter hemisphere the informational entropy values are greater than in the summer one and, which is more, the seasonal changes in H_N exceed those in H_S approximately twofold.

4.8 The concept of atmospheric vorticity charge and its applications

A quantity

$$Z_A = \iiint \tilde{\Omega} \rho d\tau \tag{1}$$

is called the total atmospheric vorticity charge. Here, integration is extended over the entire atmospheric mass. Due to Ertel's (1942) theorem, for adiabatic motion in the field of potential external forces only,

$$dZ_A/dt = \iiint (D\tilde{\Omega}/Dt)\rho d\tau = 0.$$

Taking into account that the absolute vorticity $\vec{\omega}_a$ is a solenoidal vector field, and using the Gauss–Ostrogradsky's theorem, one could write

$$Z_A = -\iint (\vec{\omega}_a \cdot \mathbf{n})(p*(\theta)/g)d\sigma.$$

Here, $d\sigma$ is an element of a surface enclosing the atmosphere, and $\mathbf{n}$ is a unit vector orthogonal to this surface and directed outward the atmospheric volume. Because the atmosphere is a spherical shell, the integration is taken over both the Earth's surface and an outer atmospheric boundary. The latter is regarded as a certain control spherical surface, co-centered with the solid Earth. It is constructed far away from the Earth's surface, i.e., 200 km above the sea level, where the fluid dynamic equations are still valid, but only a negligibly small atmospheric mass portion is dropped away. In the considered case this is less than one millionth. Due to the property $p*(\theta) \to 0$ at $\theta \to \infty$, the integral taken over the outer atmospheric boundary vanishes. A non-zero contribution is only due to integration over the Earth's surface Σ. With the account for the vector $\mathbf{n}$ directed downward, beneath the Earth, we shall have

$$Z_A = \iint_{\Sigma} \omega_{az}(p*/g)d\sigma, \qquad (2)$$

where ω_{az} is the vertical, directed upward, absolute vorticity component estimated on the Earth's surface.

In the most general case, the vorticity charge conservation law holds (Obukhov, 1962; Haynes and McIntyre, 1987, 1990; Lait, 1994)

$$\frac{\partial}{\partial t}\rho\tilde{\Omega} + \nabla \cdot \mathbf{j} = 0,$$

$$\mathbf{j} = \mathbf{v}\rho\tilde{\Omega} - \vec{\omega}_a\dot{\chi} - \chi\nabla \times \mathbf{F}, \qquad (3)$$

which excludes the existence of the vorticity charge internal sources and sinks. Here, $\mathbf{F}$ is an arbitrary non-potential force field, including friction;

$\chi = -p^*(\theta)/g$, and $\dot{\chi}$ are local χ-sources due to diabatic heating. An integral form of Equation (3), with the account for Equation (2), is of interest:

$$\frac{d}{dt}\iint_{\Sigma} \omega_{az} \frac{p^*}{g} d\sigma = \iint_{\Sigma}\left\{\omega_{az} g^{-1}\frac{dp^*}{d\theta}\dot{\theta} + (\nabla \times \mathbf{F})_z \frac{p^*}{g}\right\}d\sigma, \qquad (4)$$

As it is seen, the atmospheric vorticity charge temporal change is determined by its net flux J across the Earth's surface Σ. This flux consists of the heat J_H and frictional J_F components. With the account for both quasi-static and quasi-geostrophic nature of large-scale atmospheric motions and the atmospheric stable vertical stratification, the sign of J_H in the Northern Hemisphere is opposite to that of the diabatic heating rate $\dot{\theta}$. In the Southern Hemisphere these signs coincide. Due to this circumstance, the surface diabatic heating provokes the destruction of the positive vorticity charge accumulated in the atmosphere over the Northern Hemisphere as well as the destruction of the negative vorticity charge in the Southern Hemisphere's atmosphere. In the Northern Hemisphere, the frictional component of the vorticity charge flux is negative in cyclonic areas and positive in anticyclonic ones. In the Southern Hemisphere the pattern is reversed. Now, it is seen that a compensation, at least partial, of J_H and J_F is possible. It does happen when the surface air diabatic heating in anticyclonic areas occurs, and in cyclonic areas the diabatic cooling takes place.

In a formal way, using similar ideas proposed by Lorenz (1955) in the available potential energy theory (see Section 4.3), one might avoid the necessity to consider the vorticity charge fluxes across the Earth's surface if one smoothly extrapolates both isentropic surfaces and vorticity tubes underneath the Earth, allowing the existence of a certain 'weightless' perfect gas there. In this case, all the isentropic surfaces, among them those with $\theta < \theta^* \cong 320$ K which intersect with the Earth, become closed, and a gas beneath the Earth's surface is ascribed a certain amount of vorticity charge.[3] Now, one can speak of the vorticity charge conservation for the extended system, i.e., for the atmosphere completed with the weightless substance introduced above. The latter could be characterized conditionally

---

[3] Hoskins (1991) (see also Holton *et al.* (1995)) proposed a rational division of the atmosphere into three major parts: the 'overworld' with $\theta > 380$ K (isentropic surface $\theta \approx 380$ K corresponds roughly to the tropical tropopause position), the 'middleworld' where $\theta^* < \theta < 380$ K, which follows the property of the closeness of isentropic surfaces at $\theta \geq \theta^*$, and the 'underworld' at $\theta < \theta^*$, where isentropic surfaces lie entirely in the troposphere and intersect with the Earth's surface.

as the 'vorticity charge vacuum'. In the course of atmospheric diabatic heating, isentropic surfaces descend beneath the Earth and a corresponding portion of the atmospheric vorticity charge, for which these surfaces are impermeable (Haynes and McIntyre, 1987, 1990), is transferred to the vacuum. In contrast, when atmospheric diabatic cooling happens, the vorticity charge is extracted from the vacuum and transferred to the atmosphere. This is a rough scheme of the observed annual course in the atmospheric vorticity charge (see below).

The artificial approach used might be attempted to be laid into the basis of a method of integration of the atmospheric governing equation, written in θ-coordinates. A primary domain with curvilinear and non-stationary lower boundary is imbedded into a broader domain, containing a weightless gas beneath the Earth's surface, but enclosed from below by a fixed isentropic surface, with a simple boundary condition being imposed on it. For example, it could be a condition of the vertical velocity component vanishing $D[\theta(\partial M/\partial\theta) - M]/Dt = 0$.

We re-write the vorticity charge flux density in Equation (4) in a form more convenient for estimates:

$$j_z = \omega_{az} g^{-1} \frac{dp^*}{d\theta}\dot{\theta} + (\nabla\times\mathbf{F})_z \frac{p^*}{g}$$

$$\approx -\frac{l}{g}\frac{p^*}{\theta}\frac{g}{R(\gamma_a - \gamma)}\dot{\theta} - \Lambda(\nabla\times\mathbf{v})_z\frac{p^*}{g}.$$

An appearing constant Λ, with the dimension of the reciprocal of time, has to be determined. A condition of the annihilation of the 'frictional' and 'heat' vorticity charge fluxes $j_z = 0$ is formulated as

$$\dot{\theta}/\theta = -(\Lambda/l)R(\gamma_a - \gamma)g^{-1}(\nabla\times\mathbf{v})_z.$$

With the help of the baroclinicity parameter $\alpha^2 = R(\gamma_a - \gamma)/g \approx 0.1$, we re-write this condition in the form

$$\dot{\theta}/\theta = -(\Lambda/l)\alpha^2(\nabla\times\mathbf{v})_z.$$

To estimate Λ, a modification of the Ekman boundary layer problem (Ekman, 1905), proposed by Taylor (1914), is used. Some more details are given in Section 5.2. The following equations are solved

$$-l(v - v_g) = \nu_T\frac{\partial^2 u}{\partial z^2}, \quad l(u - u_g) = \nu_T\frac{\partial^2 v}{\partial z^2}, \tag{5}$$

where ν_T stands for the eddy viscosity, u_g and v_g are the Cartesian geostrophic wind components. A complex variable $w = u + iv$ is introduced, such that $w_g = u_g + iv_g$. In notations $\tilde{w} = w - w_g$ we have, instead of Equations (5):

$$\nu_T\left(d^2\tilde{w}/dz^2\right) - il\tilde{w} = 0.$$

This equation has a general solution

$$\tilde{w} = A\exp\left\{(1+i)\frac{z}{h}\right\} + B\exp\left\{-(1+i)\frac{z}{h}\right\},$$

where A and B are complex constants and the notation $h^2 = 2\nu_T/l$ is introduced. The main nuance is in imposing the boundary condition. By Ekman, $w \to w_g$, $\tilde{w} \to 0$ at $z \to \infty$ and $w = 0$, $\tilde{w} = -w_g$ at $z = 0$. By Taylor, $\tilde{w} \to 0$ at $z \to \infty$, as above, but a weaker slipwise condition $\partial w/\partial z = \lambda w$, $\partial\tilde{w}/\partial z = \lambda(\tilde{w} + w_g)$ is used at $z = 0$. Here, λ is a constant quantity with the dimension of the reciprocal of length. At $\lambda \to \infty$, Taylor's problem reduces to Ekman's one. A more general case of a non-linear boundary condition $\partial w/\partial z = \lambda(w)w$ has been studied in Ingel' and Mikhailova (1990). In notations $p = (\lambda h)^{-1} + 1$, $q = (\lambda h)^{-1}$ the solution of Taylor's problem could be written in the form

$$\tilde{w} = B\exp\left\{-(1+i)\frac{z}{h}\right\},$$

$$B = B_1 + B_2 = -\frac{u_g p + v_g q}{p^2 + q^2} + i\frac{u_g q - v_g p}{p^2 + q^2},$$

which is followed immediately by

$$(\nabla \times \mathbf{F})_z = \nu_T\frac{\partial^2}{\partial z^2}(\nabla \times \mathbf{v})_z = -\frac{2\nu_T}{h^2}\frac{q}{p^2 + q^2}(\nabla \times \mathbf{v})_z.$$

As the result, one has $\Lambda/l = \lambda h/[(\lambda h + 1)^2 + 1]$. The quantity Λ/l, as a function of its argument λh, reaches the maximum value $\left[2(\sqrt{2} + 1)\right]^{-1} \approx 0.207$ at $\lambda h = \sqrt{2}$. On the other hand, it is not difficult to relate λ to the angle φ between the surface and geostrophic wind. Here, a simple relation holds: $\tan\varphi = \lambda h/(\lambda h + 2)$. According to observational data,

$\varphi \approx 30°$, which corresponds to $\lambda h = 2/(\sqrt{3}-1) \approx 2.732$ and to $\Lambda/l = (1/4)(\sqrt{3}-1) \approx 0.183$, respectively. This does not deviate significantly from the estimate 0.207, resulting from the maximization Λ/l. Note that the maximum of Λ/l corresponds to $\tan\varphi = (1+\sqrt{2})^{-1}$ and $\varphi \approx 22.5°$. Assuming that Λ/l is close to its maximum value, we arrive at a simple estimate $\dot{\theta}/\theta \cong -0.02(\nabla \times \mathbf{v})_z$. For example, consider a case of an anticyclone, with relative vorticity -0.5×10^{-4} s^{-1} at a 500 hPa level, the surface relative vorticity being 40% of this value (Charney and Eliassen, 1949). Now, under assumption that surface air potential temperature is close to 300 K, we arrive at an estimate $\dot{\theta} \approx 1\,\mathrm{K} \cdot \mathrm{days}^{-1}$. It could not be excluded that this mechanism is responsible for ultra-long, up to 2 month, blocking high maintenance, provoking severe droughts in the case of summer blocking events (Agayan et al., 1986). In this particular case, favourable conditions for surface air radiative overheating are established over the anticyclonic region, which promotes the preservation of anticyclonic circulation.

In the atmospheric general circulation processes the request of identical vanishing of the vertical component of vorticity charge flux density $j_z = 0$ is excessive and unjustified. It is more natural to restrict oneself with the requirement of approximate vanishing of the net vorticity charge flux $J = J_H + J_F$ across the entire Earth's surface, or at least its hemispheric part. To be definite, consider a case of winter atmospheric circulation in the Northern Hemisphere. Here, cyclonic areas, along with surface air diabatic heating inside them, are the territories where the vorticity charge is transported away from the atmosphere. Such a process might happen to be the most extensive inside the so called 'energy active zones', i.e., within the regions over Gulfstream and Kuroshio (Atmosphere, Ocean, Space – Russian 'Razrezy' Program, 1983–1990). Areas where the vorticity charge enters the atmosphere are most likely to be less localized. Geographically, it happens where one has both anticyclonic circulation at the lower atmospheric levels and surface air diabatic cooling. Such conditions are realized over gross lands of Siberia and North America.

Formula (2) could serve as the basis for computations of the vorticity charge accumulated in the atmosphere, using the given surface potential temperature and wind velocity fields, provided a functional dependence $p^*(\theta)$ is specified. However, it is more convenient to use the possibility to determine independently Z_A on the basis of calculations of air mass distribution density $\mu(\theta,\tilde{\Omega})$ on θ and $\tilde{\Omega}$ values (see Section 4.7). Using the μ function, which is normalized by the condition

$$\iint \mu(\theta,\tilde{\Omega})\,d\theta\, d\tilde{\Omega} = 1,$$

where integration is taken over all θ and $\tilde{\Omega}$ values, the net atmospheric vorticity charge is given by the formula

$$Z_A = m_A \iint \tilde{\Omega}\mu(\theta,\tilde{\Omega})d\theta d\tilde{\Omega}$$

(m_A is the total atmospheric mass) and is related to the mass-weighted potential vorticity $\tilde{\Omega}_A$ by the formula $Z_A = \tilde{\Omega}_A m_A$. Actually, it was not the function $\mu(\theta,\tilde{\Omega})$ itself but the corresponding probability densities $\mu_N(\theta,\tilde{\Omega})$ and $\mu_S(\theta,\tilde{\Omega})$ for the Northern and Southern Hemisphere, respectively, which had been calculated in Kurgansky and Tatarskaya (1987, 1990). The equality holds

$$\mu(\theta,\tilde{\Omega}) = \frac{1}{2}\left[\mu_N(\theta,\tilde{\Omega}) + \mu_N(\theta,\tilde{\Omega})\right].$$

Here, the normalization conditions are satisfied

$$\iint \mu_N(\theta,\tilde{\Omega})d\theta d\tilde{\Omega} = \iint \mu_S(\theta,\tilde{\Omega})d\theta d\tilde{\Omega} = 1,$$

and it is assumed that $m_N = m_S = m_A/2$, where m_N and m_S are the atmospheric masses over the Northern and Southern Hemisphere, respectively. If the atmospheric vorticity charge for the Northern and Southern hemispheres is determined by the formula

$$Z_N = m_N \iint \tilde{\Omega}\mu_N(\theta,\tilde{\Omega})d\theta d\tilde{\Omega}, \quad Z_S = m_S \iint \tilde{\Omega}\mu_S(\theta,\tilde{\Omega})d\theta d\tilde{\Omega},$$

then $Z_A = Z_N + Z_S$. The corresponding $\tilde{\Omega}_N$ and $\tilde{\Omega}_S$ values are related to $\tilde{\Omega}_A$ by the formula $2\tilde{\Omega}_A = \tilde{\Omega}_N + \tilde{\Omega}_S$. Computational results are given in Table 2. As it follows from Table 2, FGGE annually mean net atmospheric vorticity charge per unit mass is close to the value

$$\frac{1}{12}\left(\frac{0.51-0.73}{2}\right)\times 10^{-4}\,\text{s}^{-1} = -0.01\times 10^{-4}\,\text{s}^{-1}.$$

Thus, the Earth's atmosphere as a whole is 'charged' negatively. This might be a manifestation of the thermal asymmetry between the hemispheres: the northern hemispheric atmosphere, if annually averaged, is slightly warmer than that of the Southern Hemisphere.

It is clear that two functions μ_N and μ_S contain more information than μ alone. Thus, the informational entropy H of the μ-distribution determined by formula (1) of Section 4.7 always exceeds the sum of informational entropies H_N and H_S for the distributions μ_N and μ_S. Remarkable empirical evidence is that nearly all air parcels with $\tilde{\Omega} > 0$ are concentrated over the Northern Hemisphere, and nearly all air parcels with $\tilde{\Omega} < 0$ over the Southern Hemisphere. That is why the functions μ_N and μ_S can be regarded as independent and their informational entropy values could be summed up: $H = H_N + H_S$. In the general case, the informational entropy difference $\Delta H = H - (H_N + H_S)$ describes the dynamic interaction (air mass exchange) between the hemispheric atmospheres.

Table 2 Monthly-averaged (FGGE period) vorticity charge values per unit atmo- spheric mass for the Northern $\tilde{\Omega}_N$ and Southern $\tilde{\Omega}_S$ Hemispheres, as well as the doubled value of the net atmospheric vorticity charge per unit mass, $2\tilde{\Omega}_A = \tilde{\Omega}_N = \tilde{\Omega}_S$.

Year, month		$\tilde{\Omega}_N$	$\tilde{\Omega}_S$	$2\tilde{\Omega}_A$	Δ^*
			$10^4 \, s^{-1}$		$10^3 \, s$
1978	XII	0.63	−0.54	0.09	2.98
1979	I	0.65	−0.52	0.13	2.99
	II	0.64	−0.52	0.12	2.84
	III	0.61	−0.52	0.09	3.05
	IV	0.56	−0.56	0.00	3.10
	V	0.50	−0.57	−0.07	2.79
1979	VI	0.43	−0.57	−0.14	2.26
	VII	0.41	−0.59	−0.18	1.92
	VIII	0.41	−0.60	−0.19	2.13
	IX	0.45	−0.58	−0.13	2.33
	X	0.53	−0.55	−0.02	2.78
	XI	0.61	−0.53	0.08	2.55

Using the thermal wind relation, it could be shown that the total atmo- spheric vorticity charge Z_A is proportional to the difference between the axial components of relative atmospheric angular momentum over the Northern Hemisphere M_N and the Southern Hemisphere M_S (Kurgansky, 1991):

$$Z_A = C(M_N - M_S). \qquad (6)$$

Here, C is a factor whose concrete value could be found in the paper just cited. Because the meteorological field seasonal changes in the Northern

Hemisphere noticeably exceed those in the Southern Hemisphere, the seasonal variations of the right-hand side of relation (6), along with those of Z_A, correlate well with the changes in the relative angular momentum of the entire atmosphere $M_A = M_N + M_S$. The latter quantity determines the seasonal variations in the total atmospheric angular momentum, with minor corrections due to the planetary (so-called 'omega') angular momentum contribution. Atmospheric angular momentum variations could be estimated independently, with the help of precise astronomical methods of determining the length-of-day fluctuations (see, e.g., Hide et $al.$, 1980; Barnes et $al.$, 1983; Bell et $al.$, 1991). The monthly-mean values of the length-of-day deviations Δ * from the standard value 86400 s for FGGE period (Bureau International de l'Heure (BIH) 1979, 1980 Rapport annuel pur 1978, 1979. Paris) are also given in Table 2.

The coefficient of correlation between the seasonal changes in H_N, H_S (see Table 1) and $\tilde{\Omega}_N$, $\tilde{\Omega}_S$ (see Table 2) is equal to $r = 0.97$, -0.87, respectively, and is meaningful in both cases. This indicates a possible functional relation between these two variables. The simplest hypothesis is to assume a Boltzmann-type distribution for the probability densities μ_N and μ_S:

$$\mu_B(\theta, \tilde{\Omega}) = \frac{A(\theta)}{B} \exp\left\{ -\frac{|\tilde{\Omega}|}{B} \right\}. \qquad (7)$$

Here, $A(\theta)$ is a weighting function, such that $\int A(\theta) d\theta = 1$ and B is a positive constant. It is easy to show that $B = \tilde{\Omega}_N$ for the Northern Hemisphere and $B = -\tilde{\Omega}_S$ for the Southern Hemisphere. Along with this, the informational entropy determined by formula (1) of Section 4.7 and the vorticity charge are functionally related

$$H_B = K \ln B + K \ln(e/A). \qquad (8)$$

Note that 'Boltzmann's' entropy defined by formula (8) is greater than the actual informational entropy H, provided the atmospheric vorticity charge value is kept constant, as well as the air mass distribution on θ values. Estimates show that the relative informational entropy 'deficit' ($H_B - H)/H_B$ is about 3–4%. This is one order of magnitude less than that of the relative informational entropy deficit, which can be estimated using both formula (2) from Section 4.7 and data from Table 1.

Below, some arguments are given, attempting to justify distribution (7) and determine the weighting function $A(\theta)$. Equation (3) does not change its form, or is covariant, under the transformation (cf. Kurgansky and Pisnichenko, 2000)

$$\chi \Rightarrow \chi^* = \Phi(\chi), \quad \tilde{\Omega} \Rightarrow \tilde{\Omega}^* = \Phi'\tilde{\Omega}.$$

Here, Φ is an arbitrary differentiable function and Φ' represents a derivative of Φ with respect to the argument χ. In Section 4.5.4 it has been shown that Friedmann's (1934) condition for $(\chi, \tilde{\Omega})$-flux tube conservation is not violated under the transformation

$$\chi \Rightarrow \chi^{**} = \Psi(\chi), \quad \tilde{\Omega} \Rightarrow \tilde{\Omega}^{**} = [\Psi']^{-1}\tilde{\Omega},$$

where Ψ is another arbitrary differentiable function of the argument χ. These two transformations are consistent in the case

$$\Phi = \Psi = \chi,$$

i.e., the Friedmann's condition might be formulated, if at all, only for a single couple of variables $(\chi, \tilde{\Omega})$. The fulfillment of Friedmann's requirements is essentially identical to the air mass constancy for every zonally oriented infinitely thin $(\chi, \tilde{\Omega})$ flux tube (cf. Kurgansky and Tatarskaya (1990) and Nakamura (1995), where a general form of continuity equation in $(\chi, \tilde{\Omega})$ coordinates is discussed) and one can speak on a stationary atmospheric mass distribution between infinitely thin $(\chi, \tilde{\Omega})$ flux tubes. Starting from the vorticity charge concept, when considering for definiteness the case of the Northern Hemisphere, and in close resemblance to (7) the following stationary reference distribution is introduced as 'an intelligent guess'

$$\mu_B(\chi, \tilde{\Omega}) = M \exp\{-\tilde{\Omega}/\tilde{\Omega}_N\}, \tag{9}$$

where $M = 1/(\tilde{\Omega}_N |X|)$, and X is the minimum χ-value, being proportional to the total hemispheric air mass. Here, air mass is equipartitioned between equal increments $d\chi$ and equal intervals of $p^*(\theta)$, as well. This is a very useful property of μ_B-distribution because it enables one to identify $p^*(\theta)$ with the average pressure on isentropic surfaces, in particular (see Section 4.5.1).

In the $(\theta, \tilde{\Omega})$ coordinates used in Kurgansky and Tatarskaya (1987), Obukhov et al. (1988) and Pisnichenko and Kurgansky (1996) the corresponding distribution

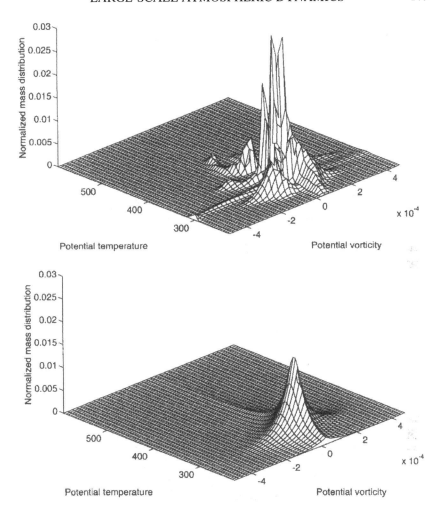

FIGURE 20 Atmospheric air mass distribution on an optimally modified PV and potential temperature θ for January 1980 (a) (from Kurgansky, M. V. and Pisnichenko, I. A. (2000) 'Modified Ertel's Potential Vorticity as a Climate Variable', *JAS* **57**, 822–835. © 2000 American Meteorological Society (AMS)); the corresponding reference Boltzmann-like distribution (b). Figure courtesy of Dr. I. A. Pisnichenko.

$$\mu_B(\theta,\tilde{\Omega}) = \left|\partial(\chi,\tilde{\Omega})/\partial(\theta,\tilde{\Omega})\right|\mu_B(\chi,\tilde{\Omega}) = \left|\chi'(\theta)\right|M\exp\left\{-\tilde{\Omega}/\tilde{\Omega}_N\right\}$$

has a maximum in the vicinity of $\theta = \theta_0$, where $|\chi'(\theta)| = \max$, which is in quantitative agreement with the plot depicted in Figure 18, see also Figure 20. It follows also that $A(\theta) \equiv |\chi'(\theta)/X|$.

Further on, we pass to the one-dimensional probability density

$$\mu_B(\tilde{\Omega}) = \tilde{\Omega}_N^{-1} \exp\{-\tilde{\Omega}/\tilde{\Omega}_N\} \tag{10}$$

by integrating (9) over all χ-values. The closeness between the actual monthly-mean $\mu(\tilde{\Omega})$-distribution and the corresponding Boltzmann-like $\mu_B(\tilde{\Omega})$-distribution has been mentioned by Kurgansky and Prikazchikov (1994), who used a PV field computed in Kurgansky and Tatarskaya (1987) and Obukhov et al. (1988), with the help of the climatological data for χ definition.

The one-dimensional probability density $\mu(\tilde{\Omega})$ of potential vorticity values for a randomly chosen air parcel, with the prescribed first moment $\tilde{\Omega}_N$ of $\mu(\tilde{\Omega})$, coincides with the probability density of vorticity charge values $\tilde{\Omega}$ for an appropriately constructed ensemble of hemispheric atmospheres, provided the expectation value of $\tilde{\Omega}$ for a given $\mu(\tilde{\Omega})$ is equal to $\tilde{\Omega}_N$. The realizations of this ensemble are the hypothetical hemispheric atmospheres, with a completely mixed, or homogenized, potential vorticity field $\tilde{\Omega}$. Here, the total hemispheric air mass is set equal to unity, which enables one to identify the hemispheric vorticity charge with $\tilde{\Omega}$. Now, the distribution (9) becomes a canonical, in Gibbs' sense, vorticity charge distribution for this ensemble. This distribution supplies the maximum of informational entropy[4]

$$H = -\int \mu \ln \mu \, d\tilde{\Omega}$$

among all other possible vorticity charge distributions between the realizations of this ensemble. The maximum value is completely determined through $\tilde{\Omega}_N$ and is equal to

$$H_B = -\int \mu_B \ln \mu_B \, d\tilde{\Omega} = \ln \tilde{\Omega}_B + \text{const.}$$

To be sure to apply distribution (10) to the case of a real atmosphere, we must adopt a set of hypotheses, which could be justified by the evidence of closeness between reference distribution (10) and that calculated on the basis of real data. The hemispheric atmosphere, with spatially non-uniform PV field, provides an example of a globally non-equilibrium, in statistical

---

[4] An example of the constructive application of the principle of informational entropy maximum (under the constraint of energy and enstrophy conservation) in the dynamic oceanography is given in Holloway (1993).

sense, system. Nevertheless, it is assumed that its almost every small portion, or subsystem, with quasi-uniform PV field, is observed in the state of local statistical equilibrium.[5] Only a small number of subsystems give the exclusion. This means that after a long enough, but still finite, time- interval the local (averaged over the sub-domain) PV-value passes through all the range of possible PV variations. In other words, a very strong mixing occurring in a PV-functional space, sometimes called K-mixing, i.e., the mixing in Kolmogorov's sense (see Dymnikov and Filatov, 1997) is assumed. In contrast, the time-interval in question should be small enough compared with the characteristic time of changes in 'external conditions', which are primarily $\tilde{\Omega}_N$ temporal variations due to the annual course of insolation. A compromise time-interval is estimated to be close to one month, which is in accordance with the averaging period traditionally used in classical climatology. The basis for the assumption of local statistical equilibrium is the observed very high degree of spatio-temporal variability in the PV field. In particular, as we have seen in Section 4.6, PV values in cyclonic regions in mid-latitudes may be as high as 8×10^{-4} s^{-1}, and in blocking anticyclones at the same latitudes they could be as low as $(0.3-0.5)\times10^{-4}$ s^{-1}. Moreover, these vortices are transient, being transported by wind. Thus, the range of possible PV variations in a fixed spatial position exceeds several times the PV mean value typical of these latitudes. An unprecedented high temporal variability of PV compared with all other known meteorological variables gives one the right to apply the approximation of local statistical equilibrium to this variable. A small spatial correlation radius in the PV field, i.e., the low cross-correlation between PV temporal changes in different positions provides the basis for assuming the existence of a very large number of effectively independent atmospheric subsystems, each of which is in the state of local statistical equilibrium. Thus, distribution (10) can be applied, as a reference, to the hemispheric atmosphere as a whole. The impact of the stratosphere, where the assumption of local statistical equilibrium might be not justified at all, is ignored because it contains a relatively small portion of the total atmospheric mass.

Globally, the atmosphere is permanently observed in the state of chaos, which serves as a background for more individually distinguishable irregular processes of different temporal and spatial scales (the so-called atmospheric circulation regimes and/or quasi-periodic global atmospheric phenomena with different periods, including quasi-biennial oscillations (QBO) with the average period about 27 months). Chaotic dynamics of the atmospheric climate system occurring on its attractor (Lorenz, 1995) could be regarded equilibrium in the sense that on the attractor we assume [and for

---

[5] This 'subsystem' concept is very close to the commonly used hydrodynamic definition of a fluid parcel (see, e.g., Brown, 1992).

certain atmospheric climate models it could be strictly proved (Dymnikov and Filatov, 1997)] the existence of a unique stationary statistical distribution, i.e., an essential invariant measure on the attractor does exist. To characterize the attractor of the atmospheric climate system in $(\chi, \tilde{\Omega})$-terms, it is proposed to use a hypothetical reference state, described by (9), to which the atmosphere closely approaches but never reaches precisely. This reference distribution describes a 'climate background noise'. From the standpoint of statistical fluid mechanics it could be regarded as a distribution corresponding to the equipartitioning of 'PV substance' (vorticity charge) between a finite, though large, number of degrees of freedom, which are effectively excited in the atmospheric climate system. The dynamics of the atmospheric climate system is not Hamiltonian, by all means, but the very existence of a stationary statistical distribution (9) for the climate system attractor, under certain additional assumptions, makes this approach appropriate.

In reality, even for the best possible choice of $(\chi, \tilde{\Omega})$-variables an actual $\mu(\tilde{\Omega})$-distribution deviates from the reference one $\mu_B(\tilde{\Omega})$. Primarily, this happens because of the discrepancies between diabatic and frictional forcings, actually present in the atmosphere, and those which satisfy the condition of 'external PV forcing of a special kind' (see Section 4.5.4). Qualitatively, the closeness between the actual and reference distributions is characterized by the informational entropy deficit

$$\Delta H = H_B - H.$$

There is a possibility to determine $(\chi, \tilde{\Omega})$-coordinates in question by minimizing ΔH with the help of a special class of 'trial' functions $\chi(\theta)$, containing that $\chi^*(\theta)$ which fits the observed atmospheric climate state. The proposed procedure has been tested by Kurgansky and Pisnichenko (1997, 2000) with the use of 1980–89 ECMWF data. The following trial functions

$$\chi(\Theta) = A[\pi/2 - \tan^{-1}(C(\theta - \theta_0))],$$

were taken, with variations given to the parameters C and θ_0 and the parameter A being determined through the condition of the total atmospheric mass constancy. It has been proven that the minimum of ΔH does exist and is achieved for relative informational entropy values $(H_B - H)/H_B$ of the order of a few tenths of a percent. The position of the minimum in the (C, θ_0)-plane is described through a linear dependence

$$\theta_0 - 292.55 = 321.4 \times (C - 0.04614),$$

found by applying the linear regression method (here, θ_0 is expressed in Kelvin (K) and C is given in K^{-1}). This holds for both extremal seasons

Atmospheric Mass Distribution on Potential Vorticity

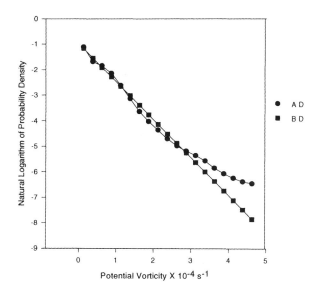

FIGURE 21 Atmospheric mass distribution on optimally modified potential vorticity for January 1980–1989, the Northern Hemisphere; AD, actual distribution and BD, the corresponding Boltzmann-like distribution. Figure courtesy of Dr. I. A. Pisnichenko.

(January and July) and for both hemispheres within the entire 10-year period. This straight line passes very close to the point $\theta_0 = 293$ K, $C = 0.04614$ K^{-1}, which accounts for $\chi^*(\theta)$, empirically found by Kurgansky and Tatarskaya (1987). Figure 21 demonstrates the evident closeness between actual $\mu(\tilde{\Omega})$ and reference $\mu_B(\tilde{\Omega})$ distribution for such an optimal PV modification. Here, the monthly-mean $\tilde{\Omega}$ statistics for January 1980–1989, the Northern Hemisphere was used. Systematic deviations occur only at large $\tilde{\Omega}$ and could be partly explained by the fact that $\tilde{\Omega}$ values are limited from above in the real atmosphere but in the framework of the $\mu_B(\tilde{\Omega})$ distribution arbitrary high $\tilde{\Omega}$ are allowed.

With the help of the mass continuity equation we re-write Equation (3) as

$$\frac{D}{Dt}\tilde{\Omega} = \frac{1}{\rho}\nabla \cdot \mathbf{N}_\chi, \quad \mathbf{N}_\chi = \vec{\omega}_a\dot{\chi} + \chi\nabla \times \mathbf{F}.$$

Obukhov (1962) has pointed out a possibility of compensation between diabatic heating and frictional PV forcing, resulting in the non-divergence of $\mathbf{N}_\chi$. As above, this property is consistent with the requirement of preservation of a covariant form of Equation (3) only for a unique choice of

$(\chi^{**}, \tilde{\Omega}^{**})$-coordinates. In these coordinates the potential vorticity $\tilde{\Omega}^{**}$ is materially conserved despite the action of diabatic heating and frictional forcing. Here, a reference stationary one-dimensional distribution (10) might be introduced, but in distinction to the previous approach, the existence of a two-dimensional reference distribution, similar to (9), remains questionable. Which of the two possible conditions for diabatic and frictional PV forcing compensation, the Friedmann's or the Obukhov's, fits the real external PV forcing in the atmosphere, remains an open problem and needs further investigations.

Simple model with exponential μ-distribution. The following arguments have been inspired by an analogy with the well-known Schmidt's (1925) explanation of the sedimentation process of dust particles and other suspensions in the turbulent atmosphere, leading to an exponential density distribution with height. Under quasi-static approximation, we may use the PV equation in the form (cf. Hoskins *et al.*, 1985)

$$\dot{\tilde{\Omega}} = \tilde{\Omega}\left(\partial\dot{\chi}/\partial\chi\right)_{\tilde{\Omega}} + f\left(\chi, \tilde{\Omega}, t\right),$$

where f describes the possible impact of frictional and latitudinal dependent diabatic PV forcing. Elimination of $(\partial\dot{\chi}/\partial\chi)_\Omega$ with the use of equation

$$\left(\partial\dot{\chi}/\partial\chi\right)_{\tilde{\Omega}} + \left(\partial\dot{\tilde{\Omega}}/\partial\tilde{\Omega}\right)_{\chi} = 0,$$

which follows from the Friedmann's (1934) necessary and sufficient conditions for the conservation of $(\chi, \tilde{\Omega})$-tubes, then integration over $\tilde{\Omega}$, and, finally, averaging of the ensemble (denoted with angular brackets) gives the Taylor–Richardson's expression for the diffusivity of air parcels in the $\tilde{\Omega}$-configuration space, for the particular choice $\tilde{\Omega} = 0$ of an 'initial' $\tilde{\Omega}$ value (cf. Monin and Yaglom, 1971, 1975; Lesieur, 1997)

$$K = \left\langle D\left(\tilde{\Omega}^2/2\right)/Dt \right\rangle = \int_0^{\tilde{\Omega}} \langle f \rangle d\tilde{\Omega}.$$

Assume that

$$c = \tilde{\Omega}\left\langle \left|\left(\partial\dot{\chi}/\partial\chi\right)_{\tilde{\Omega}}\right| \right\rangle$$

stands for the air parcel 'sedimentation rate' in the $\tilde{\Omega}$-configuration space, induced by diabatic (predominantly radiative) heating of the lowest tropospheric levels and a subsequent diabatic destruction of the PV-field (cf. Danielsen (1990) where, inversely, PV production in the stratosphere due to diabatic cooling is discussed extensively). Given an arbitrary, but fixed $\tilde{\Omega}$ level in this configuration space, the air mass flux towards low $\tilde{\Omega}$ is equal to $c\mu(\tilde{\Omega})$. In the stationary case it is cancelled by the diffusivity flux $K\,\partial\mu/\partial\tilde{\Omega}$ directed towards high $\tilde{\Omega}$:

$$K\,\partial\mu/\partial\tilde{\Omega} + c\mu = 0.$$

Integration over $\tilde{\Omega}$ gives

$$\mu = (c/K)\exp\left\{-\int_0^{\tilde{\Omega}}(c/K)d\tilde{\Omega}\right\},$$

which coincides with distribution (10) in the particular case $K/c = \tilde{\Omega}_N = $ const.

As it follows from these speculations, in the limit case $c \Rightarrow 0$, when diabatic heating of the tropospheric lowest levels is suppressed, but the surface friction strongly 'ties' the atmosphere to the underlying rotating solid Earth, one arrives at a quasi-uniform air mass distribution on PV values. In contrast, when $K \Rightarrow 0$ but the surface air heating is intensive enough, there is a tendency towards air mass concentration at zero PV values. This corresponds to the limiting case of vanishing of the differential (in latitudinal direction) atmospheric heating. Here, in the main atmospheric bulk one observes exclusively micro- and meso-scale motions with identically vanishing PV, which originate from vertical convective instability. In its major part, the first scenario is closer to the atmospheric winter-time general circulation, and the second one to the summer-time circulation, particular to that in the Northern Hemisphere. This helps us to understand more clearly the peculiarities in the seasonal course of the μ-distribution already mentioned in Section 4.7.

In the hypothetical case of $K = 0$ the hemispheric vorticity charge $\tilde{\Omega}_N = 0$ (the hemispheric air mass is set equal to unity). The work of the heat engine of atmospheric general circulation, driven by differential diabatic heating, results in the production of $\tilde{\Omega}_N > 0$. In the framework of this scheme, it is natural to assume that in order to create non-zero vorticity charge $\tilde{\Omega}_N^{(1)} + \tilde{\Omega}_N^{(2)}$ it is necessary to create, at the first step, the vorticity charge $\tilde{\Omega}_N^{(1)}$ and then the vorticity charge $\tilde{\Omega}_N^{(2)}$. If these two steps are independent, one arrives at the functional relation $\mu_B(\tilde{\Omega}_N^{(1)} + \tilde{\Omega}_N^{(2)}) = \mu_B(\tilde{\Omega}_N^{(1)}) \cdot \mu_B(\tilde{\Omega}_N^{(2)})$, which immediately leads to distribution (10).

CHAPTER 5

Dissipative Processes in the Atmosphere

For atmospheric motions with the characteristic wind speed $U = 10$ m·s^{-1} and the spatial scale $L = 10^4$ m comparable with the atmospheric scale height, the Reynolds number Re $= UL/\nu$, which in the Navier–Stokes equations characterizes the inertial-to-viscous force ratio, reaches a value of the order of 10^{10}, when the molecular air viscosity kinematic coefficient $\nu = 10^{-5}$ m^2·s^{-1} is taken. Consequently, one may disregard the molecular viscosity in the bulk of the atmosphere, except a very thin air film joining with the Earth's surface, because it is the molecular viscosity that supplies the velocity vector vanishing on the solid Earth's surface. The fact that air parcels are forced to 'stick' to the Earth's surface is the reason for an atmospheric boundary layer to appear. The latter has its own particular intrinsic dynamics and the main function to supply the needed transition from zero wind velocities at the Earth's surface to the velocities of the order of 10 m·s^{-1}, which dominate in the free atmosphere.

The concept of the atmospheric boundary layer is very much sophisticated. It is more correct to speak of an hierarchy of nested boundary layers with differentiating vertical scales, each of which has its own determining physical factors (Figure 22, see also Etling, 1996). A very thin air layer directly above the Earth's surface, where a drag force between the atmosphere and underlying surface dominates (this force exceeds all others by at least one order of magnitude) is named the surface boundary layer.

5.1 Surface boundary layer

Consider an idealized situation when a steady wind of uniform direction and with altitude-depending magnitude blows over an absolutely plane Earth's surface (see Figure 23). We choose the Cartesian axes in such a way so as

M. V. KURGANSKY

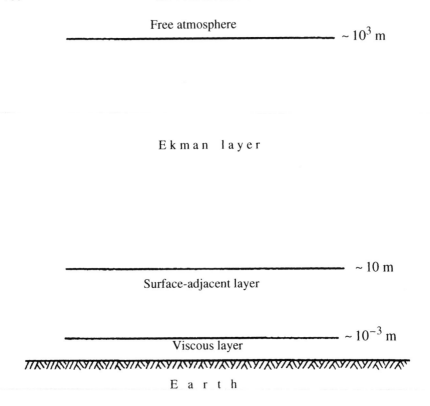

FIGURE 22   Atmospheric boundary layer structure.

to direct the *x*-axis along the wind, the *z*-axis vertically upward, and place the origin of coordinates onto the Earth's surface. We write Equation (1) of Section 1.1, being projected onto the *x*-axis

$$\rho\left(\frac{\partial u}{\partial t}+u\frac{\partial u}{\partial x}+v\frac{\partial u}{\partial y}+w\frac{\partial u}{\partial z}\right)=-\frac{\partial p}{\partial x}+\eta\nabla^2 u,$$

where the molecular viscosity dynamic coefficient η is considered as a constant. Because the surface boundary layer is very thin, and velocity of the wind is fairly small if compared with the speed of sound, air is treated as a homogeneous fluid, with $\rho =$ const, and its motion is considered to be nearly incompressible

$$\left(\partial u/\partial x\right)+\left(\partial v/\partial y\right)+\left(\partial w/\partial z\right)=0.$$

Combining these two equations, one gets

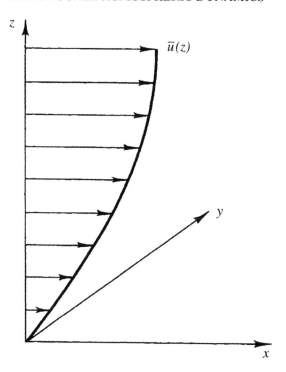

FIGURE 23   Wind profile within a surface boundary layer.

$$\frac{\partial}{\partial t}(\rho u) + \frac{\partial}{\partial x}(\rho u^2) + \frac{\partial}{\partial y}(\rho uv) + \frac{\partial}{\partial z}(\rho uw) = -\frac{\partial p}{\partial x} + \eta \nabla^2 u. \qquad (1)$$

We shall denote the running time-mean of fluid dynamic fields by a bar above, and deviations of instantaneous variable values from their means by a prime: $u = \bar{u} + u'$, $v = \bar{v} + v'$, etc. Regarding fluid dynamic fields as random, we assume their stationarity and horizontal homogeneity at every level $z = \text{const}$. After averaging Equation (1), we get

$$d\left(\overline{\rho uw}\right)/dz = \eta\, d^2\bar{u}/dz^2 \, .$$

Due to the impermeability condition $w = 0$ on the Earth's surface, $z = 0$, one can write

$$d\left(\overline{\rho u'w'}\right)/dz = \eta\, d^2\bar{u}/dz^2 \, .$$

Integration of this equation over z gives

$$\eta \, d\overline{u}/dz - \rho \overline{u'w'} = \tau_0, \tag{2}$$

where a constant τ_0 stands for the frictional stress at the Earth's surface. From the dimensional considerations it is beneficial to write $\tau_0 = \rho u_*^2$ where $u_* = 0.3$ m·s^{-1} is the so-called friction velocity. The latter numerical value has been obtained on the basis of numerous experimental data handling. With account of surface air density values close to 1.25 kg·m^{-3} we get $\tau_0 \approx 0.1$ N·m^{-2}, i.e., under typical conditions one Earth's surface square metre is affected from aside the atmosphere by a horizontal force of 0.1 N.

Our goal is to relate a vertical wind profile $\overline{u}(z)$ to the frictional stress at the Earth's surface τ_0. Due to the determining parameters of the problem we have $\overline{u}(z) = f(z; \rho, \eta, \tau_0)$. Because the dimension of velocity does not contain the mass unit, actually one gets $\overline{u} = f(z; \eta/\rho, \tau_0/\rho) = f(z; \nu, u_*)$. At first, we estimate the viscous sublayer width δ, starting from the condition that molecular viscous stresses dominate in the left-hand side of Equation (2): $\eta \, d\overline{u}/dz \approx \tau_0$. The quantity δ could be conditionally defined, for example, as such a z-value, at which turbulent stresses reach 10% of the magnitude of the viscous term (Monin and Yaglom, 1971, 1975). From dimensional arguments it follows that $\delta = C\nu/u_*$, where C is a non-dimensional constant of the order of unity. Its exact numerical value depends on the δ-definition and should be extracted from experimental data. An interval of plausible C values happens, luckily, to be not too broad. Most frequently it is assumed that $C = 5$, which for atmospheric air with $\nu = 0.15$ cm^2·s^{-1}, $u_* = 30$ cm·s^{-1} corresponds to $\delta \approx 0.3$ mm. At $z \gg \delta$ additional Reynolds stresses dominate in the left-hand side of Equation (2): $\tau_0 \approx -\rho \overline{u'w'}$. Here, because molecular viscosity does not enter any more the list of the determining parameters and we are not obligated to control the no-slip condition validity at the Earth's surface, the governing equations become Galilean invariant. Thus, the method of similarity theory and dimensional analysis has to be used to specify the vertical wind shear, which is a true Galilean invariant quantity: $d\overline{u}/dz = f(z; u_*)$ (see Landau and Lifshitz, 1988). From dimensional arguments it follows immediately that $d\overline{u}/dz = u_*/\kappa z$, where an appearing constant κ is named the von Kármán constant and is approximately equal to 0.4. The latter value has been obtained based on numerous experimental data handling. After integration, we shall have

$$\overline{u}(z_2) - \overline{u}(z_1) = (u_*/\kappa) \ln(z_2/z_1).$$

The atmospheric layer, where this relation is valid, is named the logarithmic, or Prandtl, boundary layer (Monin and Yaglom, 1971, 1975; Etling, 1996).

An underlying surface, with the vertical roughness scale not exceeding δ, is named the aerodynamically smooth surface. In nature, one usually deals with aerodynamically rough surfaces and has to introduce the roughness height z_0 to characterize them. For example, for a smooth snow field $z_0 \sim 0.1$ cm, for a grass cover $z_0 \sim 1$ cm, and for a forest $z_0 \sim 1$ m. In the case of a rough surface, the expression $d\bar{u}/dz = u_*/\kappa(z + z_0)$ is used and

$$\bar{u}(z) = \left(u_*/\kappa\right) \ln\left[\left(z + z_0\right)/z_0\right],$$

i.e., $\bar{u}(z) = 0$ at $z = 0$ and, thus, the logarithmic singularity of $\bar{u}(z)$-profile in the coordinate origin is eliminated.

By analogy with viscous stresses, following Boussinesq hypothesis, it is adopted that

$$\overline{\rho u' w'} = -\eta_T \, d\bar{u}/dz,$$

where η_T is the dynamic coefficient of turbulent (eddy) viscosity. If we introduce the kinematic eddy viscosity $v_T = \eta_T/\rho$ and take into account the above written relations, it is easy to obtain $v_T = \kappa u_* z$. A formal analogy exists between the formula just deduced and the mathematical expression for a kinematic coefficient of molecular viscosity $v \sim c\lambda$ resulting in the classical kinetic theory of gases. In the latter case, c is the characteristic velocity of chaotic molecular motion, close to the speed of sound, and λ is the free-path length of gas molecules. Under standard atmospheric conditions, $c \approx 3.2\times10^2$ m·s^{-1}, $\lambda \approx 5\times10^{-8}$ m, and $v = 0.16\times10^{-4}$ m^2·s^{-1}. On the other hand, at the height $z = 1$ m the eddy viscosity $v_T = 1.2\times10^{-1}$ m^2·s^{-1} already exceeds the molecular viscosity 10^3 times. However, one cannot consider v_T an unlimited quantity increasing with altitude. Beginning from a height of several tens of meters, the factor of stable atmospheric stratification starts to play a role. Following Obukhov (1946), let us estimate the upper limit $v_{T\lim}$, based on the Richardson's criterion for turbulence non-degeneration

$$\mathrm{Ri} = g\left(d \ln \bar{\theta}/dz\right)\Big/\left(d\bar{u}/dz\right)^2 < \mathrm{Ri}_{cr} \leq 1. \tag{3}$$

The vertical potential temperature profile $\bar{\theta}(z)$ is controlled by the heat flux across the Earth's surface. Under Boussinesq approximation, which is well-justified for this problem, it could be written (Monin and Yaglom, 1971, 1975; Landau and Lifshitz, 1988):

$$\frac{\partial}{\partial t}\left(c_p \rho \theta\right) + \frac{\partial}{\partial x}\left(c_p \rho \theta\right) + \frac{\partial}{\partial y}\left(c_p \rho v \theta\right) + \frac{\partial}{\partial z}\left(c_p \rho w \theta\right) = \lambda \nabla^2 \theta,$$

and, which is more, under the approximation in question the concepts of the potential θ and kinetic T temperature become indistinguishable. The coefficient of heat conductivity λ is assumed to be a constant quantity. Averaging this equation over time, on assumption of stationarity and horizontal homogeneity of random fluid dynamical fields, we obtain

$$\lambda d^2 \overline{\theta}/dz^2 = c_p d\left(\overline{\rho w' \theta'}\right)/dz.$$

Integrating this equation over altitude, we get

$$\lambda\left(d\overline{\theta}/dz\right) - c_p \rho \overline{w'\theta'} = -q_0,$$

where q_0 stands for the heat flux across the Earth's surface. Again, based on dimensional arguments, one could write $q_0 = c_p \rho u_* \theta_*$, where θ_* is the temperature fluctuation scale being commonly of the order of a few tenths of degree. Outside the viscous sublayer (for air, the molecular Prandtl's number is $Pr = \eta/(\lambda/c_p) = 0.7$ and, hence, the width of the thermal and viscous sublayers is nearly the same)

$$c_p \rho \overline{w'\theta'} \approx q_0.$$

Assume that $\overline{w'\theta'} = -\chi_T \, d\overline{\theta}/dz$, where χ_T is the coefficient of eddy thermal conductivity. If potential temperature increases with altitude, i.e., the atmosphere is stably stratified, then heat is transported downward, towards the Earth. Assuming that the boundary layer turbulence structure is predominantly determined by the surface friction, we take that $\chi_T \cong \nu_T = \kappa u_* z$. Now, on the basis of Equation (3) we shall have

$$-\frac{g}{\overline{\theta}} \frac{u_* \theta_*}{\kappa u_* z} \bigg/ \left(\frac{u_*}{\kappa z}\right)^2 \leq 1,$$

or

$$z \leq L = -\frac{u_*^3}{\kappa\left(g/\overline{\theta}\right)\left(q_0/c_p \rho\right)},$$

where a combination of variables L, with the dimension of the length, which appears in the right-hand side of this inequality is, as a rule, of the order of a few tens of meters. The quantity L is positive for stable atmospheric stratification, is negative for unstable stratification, when convection develops, and transits through infinity for the neutral stratification limit. In this sense, the L-scale is an analog of absolute (Kelvin) temperature in thermodynamics if positive temperatures are completed with negative ones. For the first time, the L-scale has appeared in Obukhov (1946). In a recent literature it is commonly referred to as the Monin–Obukhov length, after Monin and Obukhov (1954). Given the above, the upper limit value $v_{T\,\mathrm{lim}}$ can be re-written in the form of $v_{T\,\mathrm{lim}} = \kappa u_* L$. The Monin–Obukhov length is taken as

$$L = u_*^3 \Big/ \Big\{ \kappa \big(g/\overline{\theta} \big) \chi_{T\mathrm{lim}} \big(d\overline{\theta}/dz \big)_{\mathrm{lim}} \Big\}.$$

We use the Brunt–Vaisala frequency squared $N^2 = g\, d\ln\overline{\theta}/dz$, whose exact value corresponds to the upper limit of potential temperature gradient $\big(d\overline{\theta}/dz \big)_{\mathrm{lim}}$, and, besides, we take that $\chi_{T\mathrm{lim}} = v_{T\mathrm{lim}} = \kappa u_* L$. In this way, we arrive at the relation $L = u_*^2/\kappa^2 N^2 L$, from which it follows that $L = u^*/\kappa N$ and, consequently, $v_{T\mathrm{lim}} = u_*^2/N$. Taking $u_* = 0.3$ m·s^{-1}, $N = 10^{-2}$ s^{-1} we find that $L = 80$ m and $v_{T\mathrm{lim}} = 10$ m^2·s^{-1}. The resulting upper limit of eddy viscosity exceeds the molecular viscosity one million times.

Above the level $z \approx 0.1L$ one has to consider the deflective action of the Coriolis force.

5.2   Ekman boundary layer

Within the altitudinal range of 10–1000 m in the atmosphere the balance between the turbulent friction force and the deflective Coriolis force is the crucial factor.

An assumption of the independence of the horizontal pressure gradient components of height is made, which is common for boundary layer theories. The easiest way to demonstrate this is to take the derivatives with respect to x and y from both sides of the hydrostatic equation $\partial p/\partial z = -\rho g$. The latter equation is undoubtedly valid here, because the horizontal spatial scale of boundary layer motions significantly exceeds their vertical spatial scale. The next step is to replace the order of taking the derivatives on assumption that $\rho = \mathrm{const}$.

We start from the equations of motion

$$-lv = \frac{\partial}{\partial z} v_T \frac{\partial u}{\partial z}, \quad lu = -\frac{1}{\rho} \frac{\partial p}{\partial y} + \frac{\partial}{\partial z} v_T \frac{\partial u}{\partial z},$$

where the x-axis is directed along the geostrophic wind $u_g = -(\rho l)^{-1} \partial p/\partial y$

and the motion at every horizontal level is considered rectilinear and with uniform velocity, so that $Du/Dt = Dv/Dt = 0$, by the Galilean principle. The turbulent eddy viscosity v_T is assumed to be independent of height. The horizontal homogeneity of fluid dynamical variables is postulated. So, the two equations

$$-lv = v_T \frac{\partial^2 u}{\partial z^2}, \quad l(u - u_g) = v_T \frac{\partial^2 v}{\partial z^2} \tag{1}$$

for two unknown functions u and v hold. We multiply the second equation by the imaginary unity i, sum up the first equation, and introduce a new unknown function $w = u + iv$. As a result, we get

$$v_T (\partial^2 w / \partial z^2) = ilw - ilu_g.$$

With the help of an auxiliary variable $\tilde{w} = w - u_g$, this equation reads as

$$v_T (\partial^2 \tilde{w} / \partial z^2) - il\tilde{w} = 0.$$

It has a general solution

$$\tilde{w} = A \exp\left\{(1 + i) \frac{z}{h}\right\} + B \exp\left\{-(1 + i) \frac{z}{h}\right\},$$

where A and B are complex constants, generally speaking, and the notation $h^2 = 2v_T/l$ has been introduced for brevity. Taking $v_T = 25$ m^2·s^{-1}, which is the same order of magnitude as the value of $v_{T\lim}$ predicted in Section 5.1, and $l = 10^{-4}$ s^{-1}, we obtain $h \approx 700$ m. Due to the boundary conditions $w = 0$, $\tilde{w} = -u_g$ at $z = 0$ and $w \to u_g$, $\tilde{w} \to 0$ at $z \to \infty$, we write

$$\tilde{w} = -u_g \exp\left\{-(1 + i)(z/h)\right\}.$$

Separating the real and imaginary part of this equality, we finally have

$$u = u_g \left(1 - e^{-z/h} \cos\frac{z}{h}\right), \quad v = u_g e^{-z/h} \sin\frac{z}{h}. \tag{2}$$

This wind distribution with height can be displayed with the help of vectorial arrows beginning in the sole point and depicting the wind speed at

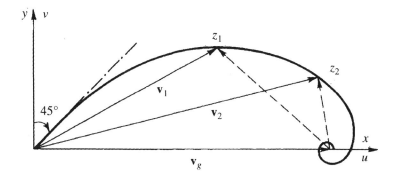

FIGURE 24 Ekman spiral over the solid Earth surface. Wind speed at the surface level vanishes, and the deflection angle is equal to 45°. At high altitudes, wind $\mathbf{v}$ becomes geostrophic $\mathbf{v}_g$. The sketch is drawn for the Northern Hemisphere. Ageostrophic wind $\mathbf{v}_a = \mathbf{v} - \mathbf{v}_g$ is shown with dashed arrows. Ekman spiral is the so called equiangular helix: in all its points an angle between $\mathbf{v}_a$ and the corresponding tangent vector is constant and is equal to 45° in the case considered.

different levels by their direction and magnitude (see Figure 24). Connecting the vector edges, we arrive at a continuous curve. What is needed more is only to draw altitudinal marks in several of its points to be able to distinguish the magnitude and direction of wind at any desired level. This curve is actually a logarithmic helix and is named the Ekman spiral after Ekman (1905) who was the first among geophysicists to arrive at the solution (2) and to construct the corresponding wind hodograph. The Ekman spiral enters the origin of coordinates in the (u,v)-plane, having an angle of 45° with the u-axis and, what is more, $u \cong v \approx u_g(z/h)$ at $z \rightarrow 0$, and approaches asymptotically $u = u_g$, $v = 0$, at $z \rightarrow \infty$. The helix intersects the u-axis an infinite number of times and performs it for such u-values which are either greater or smaller than u_g, but which are the closer to u_g the higher altitude z is. If we disregard these minor deviations from u_g, the schematic Figure 24 well agrees with the observations which also show the wind speed increase with height up to geostrophic wind values and the right wind rotation. In meteorology, the wind rotation with height shown in Figure 24 is termed the right rotation, because an observer staying with his face to the wind discovers that the wind at an upper layer blows from his right-hand side (Koschmieder, 1933).

A characterization of such wind rotation with height is the helicity $\chi = \left(\nabla \times \mathbf{v} + 4\vec{\Omega}\right) \cdot \mathbf{v}$ (see Chapter 1). For the Ekman spiral, in the above chosen right-handed coordinate system, one has

$$\chi = -u\left(\partial v/\partial z\right) + v\left(\partial u/\partial z\right),$$

where u and v are determined through expressions (2). The distribution of χ with altitude is given by the formula (see also Hide, 1989)

$$\chi = u_g^2 h^{-1} \left\{ e^{-z/h} \left(\sin \frac{z}{h} - \cos \frac{z}{h} \right) + e^{-2z/h} \right\},$$

i.e. one has the right wind rotation in the main bulk of the Ekman boundary layer, which at high altitudes is replaced by alternations of right and left rotations along with the wind vector $\mathbf{v} = (u, v)$ performing small oscillations about the vector $\mathbf{v}_g$. The total helicity in a vertical air column of unit horizontal cross-section is determined by the easily calculated integral

$$H = \int_0^\infty \chi dz = \frac{u_g^2}{2}. \tag{3}$$

However, the expression for H could be arrived at even in an easier way, if we start from the readily checked identity, valid at $v_T = \mathrm{const}$,

$$-l\chi = \frac{\partial}{\partial z} \left\{ \frac{v_T}{2} \left[\left(\frac{\partial u}{\partial z} \right)^2 + \left(\frac{\partial v}{\partial z} \right)^2 \right] + l u_g v \right\}.$$

With its use and account for $h^2 = 2v_T/l$ and $u \cong v \approx u_g(z/h)$ at $z \to 0$, we arrive at formula (3).

As it follows from the general helicity balance Equation (1) from Section 1.4, applied to the Ekman boundary layer problem, the time rate of helicity dissipation due to eddy viscosity is determined by the formula

$$2\vec{\omega}_a \cdot \mathbf{F} = -2 \frac{\partial v}{\partial z} \frac{\partial}{\partial z} v_T(z) \frac{\partial u}{\partial z} + 2 \frac{\partial u}{\partial z} \frac{\partial}{\partial z} v_T(z) \frac{\partial v}{\partial z}$$

$$= 2 \frac{\partial v}{\partial z} lv + 2 \frac{\partial u}{\partial z} l(u - u_g) = 2l \frac{\partial}{\partial z} \frac{(u - u_g)^2 + v^2}{2},$$

which is valid for an arbitrary functional dependence $v_T(z)$. With the account for the boundary conditions, we have

$$\int_0^\infty 2\vec{\omega}_a \cdot \mathbf{F} dz = -2l \frac{u_g^2}{2}.$$

So, in the absence of a divergent term in the right-hand side of the helicity balance equation the total helicity would decay in time following the exponent

$$H = \left(u_g^2 / 2 \right) \propto \exp\{-2lt\}.$$

However, it will not in fact contradict an initial assumption on the stationarity of Ekman solution (2), because the destruction of helicity is exactly compensated for by its supply from the free atmosphere (Kurgansky, 1989).

In the Southern Hemisphere, the Ekman spiral is just a mirror reflection of the curve in Figure 24 with respect to the u-axis. Correspondingly, one has to reverse the sign of both helicity H and the time rate of its destruction, assuming that in the above formula $l > 0$, thus taking the absolute value of the Coriolis parameter.

The described simple model reproduces the situation within a boundary layer qualitatively correctly, but the observed angles φ of the surface wind deviation from the geostrophic wind are smaller than the theoretically predicted value $\varphi = 45°$. Typically, the φ values are around 20–30°. For an unstable stratified boundary layer, i.e., in the case of convection, φ is rather small, about several degrees. In the case of stable stratification this angle may reach 40°. If one stays on positions of the semi-empirical turbulence theory and uses the eddy viscosity, one could attempt to improve the shortcoming of the Ekman solution by using at least two methods. The first method is to try a more realistic altitude dependence in the eddy viscosity $\nu_T = \nu_T(z)$ (Brown, 1972). The other method proposed by Taylor (1914) is to replace the non-slip condition $w = u + iv = 0$ on the Earth's surface by the free slip condition $\partial w / \partial z = \lambda w$, with λ as a certain constant. Here, the existence of the surface boundary layer is taken into account implicitly and the lower boundary condition is posed, actually, at its top, say, at a height $z = 10$ m. When $\lambda \to \infty$, Taylor's problem reduces to Ekman's problem. Algebraic transformations are a little bit more cumbersome in Taylor's problem, but one manages to link the parameter λ with the wind deviation angle φ by a simple relation $\tan\varphi = \lambda h / (2 + \lambda h)$.

If τ_x, τ_y are the components of the friction force, acting onto a unit Earth's surface area from aside the atmosphere, then according to the third Newton's law

$$\tau_x + i\tau_y = -\int_0^\infty \rho \frac{\partial}{\partial z} \nu_T \frac{\partial w}{\partial z} dz = \rho \nu_T \frac{\partial w}{\partial z}\bigg|_{z=0}.$$

Due to the free slip condition on the Earth's surface

$$\tau_x + i\tau_y = \rho v_T \lambda w\big|_{z=0} = \rho v_T \lambda (u + iv)\big|_{z=0}$$

and the direction of the vector of surface friction stresses $\vec{\tau} = (\tau_x, \tau_y)$ coincides with that of the surface wind $\mathbf{v}_s = (u_s, v_s)$, deviating from the direction of the geostrophic wind $\mathbf{v}_g = (u_g, v_g)$ by an angle φ. Expressing λ in terms of this angle, we shall have

$$\vec{\tau} = \frac{2\rho v_T}{h} \frac{\sin\varphi}{\cos\varphi - \sin\varphi} \mathbf{v}_s. \tag{4}$$

Taylor (1914), based on both general fluid dynamic arguments and the dimension theory, suggested the validity of quadratic, or ballistic, resistance law

$$\vec{\tau} = c_D \rho |\mathbf{v}_s| \mathbf{v}_s, \tag{5}$$

where by experimental evidence $c_D \sim 10^{-3}$. Currently, formula (5) is widely used in numerical modeling of large-scale atmospheric processes (Lorenz, 1967; Van Mieghem, 1973).

Comparison of formulae (4, 5) gives that

$$(2v_T/h)\sin\varphi/(\cos\varphi - \sin\varphi) = c_D |\mathbf{v}_s|.$$

The modulus of the surface wind vector $|\mathbf{v}_s|$ is related to the geostrophic wind vector modulus $|\mathbf{v}_g|$ as $|\mathbf{v}_s| = |\mathbf{v}_g|(\cos\varphi - \sin\varphi)$. Keeping in mind that $2v_T = h^2 l$, we finally get

$$hl\sin\varphi/(\cos\varphi - \sin\varphi)^2 = c_D |\mathbf{v}_g|.$$

Taking $\varphi = \pi/8$, $l = 10^{-4}\,\mathrm{s}^{-1}$, $c_D = 0.003$, $|\mathbf{v}_g| = 10\,\mathrm{m\cdot s^{-1}}$, we shall have $h \approx 230\,\mathrm{m}$ and $v_T \approx 2.7\,\mathrm{m^2 \cdot s^{-1}}$. The latter value is somewhat lower if compared with our previous estimates. This gives evidence to the uncertainty level which is immanent to this sort of boundary layer theories.

5.3   Ekman friction mechanism

Assume that the geostrophic wind velocity depends weakly (or, as it is sometimes said, parametrically) on the transversal, with respect to the wind direction, coordinate y: $u_g = u_g(y)$. Let us calculate the appearing vertical velocity at the top of the Ekman boundary layer. We start from the incompressibility equation

$$\left(\partial u/\partial x\right) + \left(\partial v/\partial y\right) + \left(\partial w/\partial z\right) = 0.$$

When integrating it over height, one gets

$$w\big|_{z\to\infty} = -\int_0^\infty \left(\frac{\partial u}{\partial x} + \frac{\partial v}{\partial y}\right) dz.$$

Now, with account for Equations (1) from Section 5.2, we shall have

$$w\big|_{z\to\infty} = -\frac{v_T}{l}\int_0^\infty \frac{\partial^2}{\partial z^2}\left(\frac{\partial v}{\partial x} - \frac{\partial u}{\partial y}\right) dz =$$

$$= \frac{v_T}{l}\frac{\partial}{\partial z}\left(\frac{\partial v}{\partial x} - \frac{\partial u}{\partial y}\right)\bigg|_{z=0} = \frac{1}{\rho l}\left(\frac{\partial \tau_y}{\partial x} - \frac{\partial \tau_x}{\partial y}\right),$$

where τ_x and τ_y stand for surface friction stress components. According to the Ekman solution $\tau_x = \tau_y = \rho v_T u_g/h$, and thus

$$w\big|_{z\to\infty} = -\left(h/2\right) du_g/dy.$$

If we introduce a relative vorticity vertical component in the free atmosphere, the formula obtained, which has been derived by Dubuc (1947) and Charney and Eliassen (1949), could be written in an invariant form

$$w\big|_{z\to\infty} = w_E = \left(h/2\right)\nabla^2\psi, \tag{1}$$

where ψ stands for the geostrophic stream function.

One might say that some sort of a pump works inside the Ekman boundary layer. In a cyclonic region, air is pumped aloft, into the free atmosphere. The other way round, where anticyclonicity occurs, it is sucked into the boundary layer. Such an asymmetric behavior of vertical velocity actually represents an adequate reaction of boundary layer intrinsic dynamics to large-scale motion in the free atmosphere, which is generally characteristic of stable systems (cf. Le Chatelieur's principle in thermodynamics or Lentz' rule in electrodynamics). As the Ekman spiral shows, the wind is always deviated towards the lower pressure. In a cyclonic region, this results in a convergence of the mass flux which, due to a weak air compressibility, leads to updraft motions developing at the top of the boundary layer. In contrast, inside an anticyclone, a divergence of the mass flux occurs and the vertical velocity at the top of the Ekman boundary layer is directed downwards. In the former case, a free atmospheric effect of vortex tube contraction in the vertical direction is observed. This, due to the theorem of potential vorticity conservation, results in the damping of the vertical component of absolute vorticity. In the latter case, the tubes are stretching, vorticity increases and an anticyclone weakens.

Addressing a more formalistic treatment of the Ekman friction mechanism, we start from the quasi-geostrophic potential vorticity equation (see Equation (5) of Section 2.3)

$$\frac{D}{Dt}\left(\nabla^2\psi + l + \frac{\partial}{\partial p}m^2 p^2 \frac{\partial\psi}{\partial p}\right) = 0. \qquad (2)$$

Relation (1) supplies the lower boundary condition for Equation (2); what is more, one might write without the loss of accuracy (because the quasi-geostrophic approximation is valid)

$$w = \left(\frac{\partial}{\partial t} + u\frac{\partial}{\partial x} + v\frac{\partial}{\partial y} + w\frac{\partial}{\partial z}\right)z = w_E, \quad z = 0,$$

and in p-coordinates, respectively,

$$\frac{Dz}{Dt} + w*\frac{\partial z}{\partial p} = w_E, \quad p = p_0.$$

Combining these relations with the thermodynamic equation, we shall have for geostrophic stream function $\psi = gz/l$ that

$$\frac{D}{Dt}\left(p\frac{\partial\psi}{\partial p} + \alpha^2\psi\right) = \frac{\alpha^2 gh}{2\tilde{l}}\nabla^2\psi, \quad p = p_0. \qquad (3)$$

We perform a reduction to the barotropic atmosphere case by integrating Equation (2) over pressure with the account for the boundary condition (3) and the corresponding condition at the atmospheric top (see Chapter 2):

$$\int_0^{p_0}\frac{D}{Dt}\left(\nabla^2\psi + l\right)dp + \int_0^{p_0}\frac{\partial}{\partial p}\frac{D}{Dt}m^2 p^2\frac{\partial\psi}{\partial p}dp = 0.$$

Utilizing the barotropic motion properties, we have commutated the derivatives D/Dt and $\partial/\partial p$ in the integrand of the second left-hand side term. After performing non-laborious transformations, we arrive at the Charney–Obukhov Equation (4) from Section 2.2 with the account for Ekman friction

$$\frac{D}{Dt}\left(\nabla^2\psi - \frac{1}{L_0^2}\psi + l\right) = -\mu\nabla^2\psi. \tag{4}$$

An appearing constant $\mu = \tilde{l}\,h/2H = gh/2\tilde{l}\,L_0^2$ has the dimension of an inverse of time (here, H is the atmospheric scale-height). Parameter $\tau = \mu^{-1}$ specifies the characteristic time-scale of relative vorticity decay due to the Ekman friction. In the atmosphere, $\tau \approx 2{\cdot}10^5$ s ≈ 2.3 days. To characterize the Ekman dissipation, the non-dimensional Ekman number $E = (h/H)^2 \approx 10^{-2}$ is sometimes used. It is easily seen that $\mu = 0.5\tilde{l}\,E^{1/2}$. On the other hand, recalling that $h^2 = 2\nu_T/\tilde{l}$, we get $E = 2\nu_T/\tilde{l}\,H^2$, i.e., under free atmospheric conditions, the Ekman number describes the ratio of the turbulent friction force to the Coriolis force.

On the basis of Equation (4), the energy balance equation is readily written

$$\frac{d}{dt}\frac{1}{2}\iint\left\{(\nabla\psi)^2 + \frac{1}{L_0^2}\psi^2\right\}dxdy = -\iint\mu(\nabla\psi)^2\,dxdy.$$

The specific rate of the energy frictional dissipation is characterized by the value

$$\varepsilon = 0.5{\times}10^{-5}\,\text{s}^{-1}\cdot 10^2\,\text{m}^2{\cdot}\text{s}^{-2} = 0.5{\times}10^{-3}\,\text{m}^2{\cdot}\text{s}^{-3},$$

which is close to empirical estimates by Oort (1964, 1983), Wiin-Nielsen and Chen (1993) and others and stresses the particular role which the Ekman layer plays to establish the atmospheric energy balance.

As for the integral of one-half of potential vorticity $\nabla^2\psi + l - \psi/L_0^2$ squared taken over the atmospheric volume, which is named the potential enstrophy, the balance of this quantity could be provided only at the expense of the horizontal, or Newtonian, viscosity, taking into account the term $\nu\nabla^4\psi$ in the right-hand side of Equation (4). The viscous dissipation of potential enstrophy occurs at the finest spatial scales λ and is determined by the integral $\iint\nu\left(\nabla^3\psi\right)^2 dxdy$. This dissipative function density ε_1, with the units of s^{-3}, completely determines the energy distribution over the wave number k spectrum within the inertial range $L^{-1} \ll k \ll \lambda^{-1} \approx \varepsilon_1^{1/6}\nu^{-1/2}$ according to the law

$$E(k) = C\varepsilon_1^{2/3} k^{-3}, \tag{5}$$

which is different from the Kolmogorov–Obukhov 'minus 5/3' power law (see Monin and Yaglom, 1971, 1975; Landau and Lifshitz, 1988). Here, L is the external scale, at which the potential enstrophy is supplied, and C is a universal non-dimensional constant. In meteorology, the law (5), generally valid for two-dimensional and quasi-two-dimensional flows (see Charney, 1971; Monin, 1990 and references therein) has found a support both in observational data and in the numerical modeling results for the zonal wave number range $k = 7$–20, which borders upon the baroclinic instability range $k = 6$–7.

5.4 Turbulent Ekman layer

Following the ideas by Dolzhanskii and Manin (1993), at the very end of this Chapter we attempt to relate the friction stress vector $\vec{\tau}$ to the geostrophic wind vector $\mathbf{v}_g$, based solely on both primary symmetry principles and basic fluid dynamic similarity arguments, but not using the eddy viscosity concept explicitly.

In the most general case, one might assume that the variables $w = u + iv$ and $w_g = u_g + iv_g$ of Section 5.2 are linearly related

$$w = Aw_g, \tag{1}$$

where $A = a + ib$, with a and b as functions of the non-dimensional vertical coordinate $\zeta = z/L$. Here, L is a characteristic scale-height of the boundary layer, which, generally speaking, depends on horizontal coordinates x, y. Below, L will be defined more precisely.

For the following purposes, it is convenient to re-write Equation (1) in tensorial notations, by using components v_j, v_{gj}($j = 1, 2$) of two-dimensional vectors $\mathbf{v} = (u,v)$, $\mathbf{v}_g = (u_g, v_g)$:

$$v_j = A_{jk}(\zeta)v_{gk},$$

where matrix A_{jk} has the form ($x_1 = x$, $x_2 = y$, $x_3 = z$)

$$A_{jk} = a\delta_{jk} + be_{3jk}.$$

Here δ_{pq} and e_{pqr} are the Kronecker and Levi–Civita tensors, respectively.

Due to the adjustment between actual and geostrophic wind at $\zeta \to \infty$, it follows that $a \to 1$, $b \to 0$, $A_{jk} \to \delta_{jk}$. Thus, it is convenient to write

$$v_j = \delta_{jk} v_{gk} + \left(A_{jk} - \delta_{jk} \right) v_{gk}. \tag{2}$$

The vertical velocity at the top of the Ekman boundary layer is given by the formula (see Section 5.3)

$$w_E = -\int_0^\infty \left(\partial v_j / \partial x_j \right) dz. \tag{3}$$

After substituting expression (2) into formula (3), we shall have

$$w_E = -\int_0^\infty \frac{\partial v_{gj}}{\partial x_j} dz - \int_0^\infty \left(A_{jk} - \delta_{jk} \right) \frac{\partial v_{gk}}{\partial x_j} dz$$

$$-\int_0^\infty \frac{d}{d\zeta} \left(A_{jk} - \delta_{jk} \right) \frac{\partial \zeta}{\partial x_j} v_{gk} dz.$$

We take into account that $\zeta = z/L$, $\partial\zeta/\partial x_j = -\zeta \, \partial \ln L / \partial x_j$ and get

$$w_E = -\int_0^\infty \frac{\partial v_{gj}}{\partial x_j} dz - \int_0^\infty \left(A_{jk} - \delta_{jk} \right) \frac{\partial v_{gk}}{\partial x_j} dz$$

$$+\int_0^\infty v_{gk} \frac{\partial}{\partial x_j} \ln L \zeta \frac{d}{d\zeta} \left(A_{jk} - \delta_{jk} \right) dz.$$

Finally, we integrate over ζ instead of z:

$$w_E = -\int_0^\infty \frac{\partial v_{gj}}{\partial x_j} L d\zeta - \int_0^\infty \left(A_{jk} - \delta_{jk} \right) \frac{\partial v_{gk}}{\partial x_j} L d\zeta$$

$$+\int_0^\infty v_{gk} \frac{\partial L}{\partial x_j} \zeta \frac{d}{d\zeta} \left(A_{jk} - \delta_{jk} \right) d\zeta.$$

Assume that the geostrophic wind is non-divergent $\partial v_{gi}/\partial x_j = 0$. After integration by parts, with the account for $A_{jk} \to \delta_{jk}$ at $\zeta \to \infty$, we get

$$w_E = -\frac{\partial v_{gk}}{\partial x_j} L \int_0^\infty \left(A_{jk} - \delta_{jk} \right) d\zeta - v_{gk} \frac{\partial L}{\partial x_j} \int_0^\infty \left(A_{jk} - \delta_{jk} \right) d\zeta.$$

In notations $B_{jk} = \int_0^\infty \left(\delta_{jk} - A_{jk} \right) d\zeta$, it finally reads as

$$w_E = B_{jk} \, \partial\!\left(v_{gk} L \right) \big/ \partial x_j , \qquad (4)$$

and, what is more, $B_{jk} = \alpha\delta_{jk} + \beta e_{3gk}$ with $\alpha = \int_0^\infty (1 - a) d\zeta$ and $\beta = \int_0^\infty b \, d\zeta$.

The constants α and β are positive, because the wind vector has the right rotation with height (see Figure 23) and its modulus $|\mathbf{v}|$ is smaller than the geostrophic wind modulus $|\mathbf{v}_g|$ almost everywhere.

Consider several particular cases of the application of the general formula (4).

In the simplest case of $L = h/2\beta = \text{const}$ we arrive at the Dubuc–Charney–Eliassen's formula, see Equation (1) of Section 5.3. Thus, the linear Ekman friction law is obtained.

Assuming $L = |\mathbf{v}_g|/l \approx 10^5$ m, we arrive at a quadratic dependence

$$w_E = l^{-1} B_{jk} \, \partial\!\left(|\mathbf{v}_g| v_{gk} \right) \big/ \partial x_j .$$

It is convenient to re-write this formula as

$$w_E = l^{-1}\alpha\nabla \cdot \left(|\mathbf{v}_g| \mathbf{v}_g \right) + l^{-1}\beta\nabla \times \left(|\mathbf{v}_g| \mathbf{v}_g \right),$$

where ∇ is the two-dimensional, in (x, y)-plane, Hamiltonian operator. With the use of the identity

$$\nabla \cdot \left(|\mathbf{v}_g| \mathbf{v}_g \right) = \nabla \times \left(\mathbf{k} \times |\mathbf{v}_g| \mathbf{v}_g \right),$$

with a unit vector $\mathbf{k}$ directed upward, we shall get

$$w_E = l^{-1}\nabla \times \left\{ \alpha\mathbf{k} \times |\mathbf{v}_g| \mathbf{v}_g + \beta|\mathbf{v}_g| \mathbf{v}_g \right\}. \qquad (5)$$

Comparing expression (5) with the general equation (see Section 5.3)

$$w_E = \rho^{-1} l^{-1} \nabla \times \vec{\tau}$$

one gets

$$\vec{\tau} = \rho\beta |v_g| v_g + \rho\alpha \mathbf{k} \times |v_g| v_g + \nabla\Phi, \qquad (6)$$

where Φ is an arbitrary scalar differentiable function, which may be set equal to zero without the loss of generality. The second term, appearing in the right-hand side of Equation (6) and absent in formula (5) of Section 5.2, is not accidental, because vectors v_s and v_g are not collinear. This term does not influence atmospheric energetics, but transforms the stability properties of large-scale atmospheric motions affected by friction (Danilov, 1992). Relation (6) could be derived directly, if we start from formula (5) of Section 5.2 and use the decomposition of vector $|v_s| v_s$ into the system of two orthogonal vectors $|v_g| v_g$ and $\mathbf{k} \times |v_g| v_g$ (in our case all the three vectors in question are complanar)

$$|v_s| v_s = C_1 |v_g| v_g + C_2 \left(\mathbf{k} \times |v_g| v_g \right).$$

Here C_1 and C_2 are certain positive constants. This is guaranteed by the right rotation of vector v with height (see Figure 24).

Charney–Obukhov's equation with the quadratic friction term in its right-hand side is written in the form (Dolzhanskii and Manin, 1993)

$$\frac{\partial}{\partial t}\left(\nabla^2\psi - \frac{1}{L_0^2}\psi \right) + J\left(\psi, \nabla^2\psi + l + \frac{\alpha}{H}|\nabla\psi| \right) = -\frac{\beta}{H}\nabla\left(|\nabla\psi|\nabla\psi\right), \qquad (7)$$

i.e., friction does not lead only to energy dissipation but redistributes energy over spatial scales, thus modifying the beta-effect. Here, as usually, J denotes the Jacobian operator.

Instead of the Coriolis parameter l, one may adopt the averaged (characteristic) value of bulk helicity density within the Ekman boundary layer χ_0 as one of the key parameters which, along with $|v_g|$, determines the scale L. The χ_0-value is determined not only by l, but also by the vertical stratification of the atmospheric boundary layer and the underlying surface roughness. As it follows from the Richardson's criterion, the second factor sets a definite upper limit to vertical wind shear. The latter factor determines the surface wind strength. For the Ekman boundary layers $\chi_0 \sim 10^{-1}$ m·s^{-2}.

Now, the characteristic vertical scale, which is determined by $L = \mathbf{v}_g^2/\chi_0$, acquires the value ~$10^3$ m, and thus happens to be of the same order of magnitude as the observed vertical scale of the boundary layer. As the result, we arrive at the cubic friction law

$$w_E = \chi_0^{-1} B_{jk}\, \partial\!\left(\mathbf{v}_g^2 v_{gk}\right)\!\big/\partial x_j \,,$$

with matrix elements B_{jk} being of the order of unity. For the cubic friction, Charney–Obukhov's equation takes the form

$$\frac{\partial}{\partial t}\left(\nabla^2\psi - \frac{1}{L_0^2}\psi\right) + J\left(\psi, \nabla^2\psi + l + \frac{\alpha\tilde{l}}{H\chi_0}(\nabla\psi)^2\right) = -\frac{\beta\tilde{l}}{H\chi_0}\nabla\!\left((\nabla\psi)^2\nabla\psi\right). \quad (8)$$

There exists the only possibility to answer the question, which of the viscous resistance laws, linear, quadratic or, maybe, cubic, suits the turbulent Ekman boundary layer conditions the best. It is to continue work with Equation (4) of Section 5.3, and Equations (7, 8), accompanying it with analysis and interpretation of experimental data.

REFERENCES

Abarbanel, H. D. I., D. D. Holm, J. E. Marsden and T. S. Ratiu (1986) Nonlinear stability analysis of stratified fluid equilibria. *Phil. Trans. R. Soc. London* **A318(1543)**, 349–409.

Agayan, G. M., M. V. Kurgansky and I. A. Pisnichenko (1986) Analysis of blocking situations in the atmosphere using adiabatic invariants and FGGE data. *Izvestiya, Atmospheric Oceanic Physics* **22(11)**, 875–880 (English translation).

Agayan, G. M., M. V. Kurgansky and I. A. Pisnichenko (1990) Calculation of a field of heat influxes using the equation of potential vorticity transformation. *Meteorologiya i Gydrologiya* **7**, 19–27 (in Russian).

Alishayev, D. M. (1980) Dynamics of the two-dimensional baroclinic atmosphere. *Izvestiya, Atmospheric Oceanic Physics* **16**, 63–67 (English translation).

Arnol'd, V. I. (1965) Conditions for nonlinear stability of stationary plane curvilinear flows of an ideal fluid. *Dokl. Akad. Nauk SSSR* **162(5)**, 975–978 (English translation 1965: *Sov. Math.,* **6**, 773–777).

Arnol'd, V. I. (1978) *Dopolnitel'nye Glavy Teorii Obyknovennykh Differentsial'nykh Uravneniy (Supplementary Chapters of the Theory of Ordinary Differential Equations)*. Nauka Publishers, Moscow, 304 pp. (in Russian).

Atkinson, B.W. (1989) *Meso-Scale Atmospheric Circulations*. Academic Press, 495 pp.

Atmosphere, Ocean, Space – Russian 'Razrezy' Program (1983–1990) *Itogi Nauki i Tekhniki VINITI Acad. Sci. USSR*, Moscow (Ed. G. I. Marchuk), **1–13** (in Russian).

Barnes, R. T. H., R. Hide, A. A. White and C. A. Wilson (1983) Atmospheric angular momentum fluctuations, length-of-day changes and polar motion. *Proc. R. Soc. London* **A387**, 31–73.

Batchelor, G. K. (1967) *An Introduction to Fluid Dynamics*. Cambridge University Press, 615 pp.

Bell, M. (1994) Oscillations in the equatorial component of the atmosphere's angular momentum and torque on the Earth's bulge. *Quart. J. Roy. Meteorol. Soc.* **120**, 195–213.

Bell, M. J., R. Hide and G. Sakellarides (1991) Atmospheric angular momentum forecasts as novel tests of global numerical weather prediction models. *Phil. Trans. Roy. Soc. London* **A334**, 55–92.

Bell, M. J. and A. A. White (1988) Spurious stability and instability in N-level quasi-geostrophic models. *J. Atmos. Sci.* **45(11)**, 1731–1738.

Blinova, E. N. (1943) Hydrodynamic theory of pressure waves and centers of atmospheric action. *Dokl. Akad. Nauk SSSR* **39(7)**, 284–287 (in Russian).

Blinova, E. N. (1946) On the determination of the velocity of motion of lows from the nonlinear vorticity equation. *Prikl. Mat. Mekh.* **X**, 5–6 (in Russian).

Bluestein, H. B. (1992) *Synoptic-Dynamic Meteorology in Midlatitudes. Vol. I: Principles of Kinematics and Dynamics.* Oxford University Press, Oxford, 426 pp.

Blumen, W. (1968) On the stability of quasi-geostrophic flow. *J. Atmos. Sci.* **25(5)**, 929–931.

Blumen, W. (1981) The geostrophic coordinate transformation. *J. Atmos. Sci.* **38**, 1100–1105.

Blumen, W. (1989) Constraints on solution of Long's equation for steady, two-dimensional, hydrostatic flow over a ridge. *J. Atmos. Sci.* **46(10)**, 1428–1433.

Bolin, B. (1955) Numerical forecasting with the barotropic model. *Tellus* **7(1)**, 27–49.

Bolin, B. (1999) Carl-Gustaf Rossby. The Stochholm period 1947–1957. *Tellus* **51 A–B**, 4–12.

Boubnov, B. M. and G. S. Golitsyn (1995) *Convection in Rotating Fluids.* Kluwer Academic Publishers, 224 pp.

Brown, R. A. (1972) *Analytical Methods in Planetary Boundary Layer Modelling.* Adam Hilger, London.

Brown, R. A. (1992) *Fluid Mechanics of the Atmosphere.* Academic Press, London.

Brunt, D. (1941) *Physical and Dynamical Meteorology.* Cambridge University Press, 428 pp.

Bugaev, V. A. (1947) *Tekhnika Sinopticheskogo Analiza i Prognoza (Technique of Synoptic Analysis and Prediction).* Leningrad, Gidrometeoizdat (Chapter 9: Isentropic analysis, 197–262) (in Russian).

Bureau international de l'heure (BIH) (1979, 1980) *Rapport annuel pur 1978, 1979. Paris.*

Butchart, N. and E. E. Remsberg (1986) The area of the stratosheric polar vortex as a diagnostic for tracer transport on an isentropic surface. *J. Atmos. Sci.* **43(13)**, 1319–1339.

Cahn, A., Jr. (1945) An investigation of the free oscillations of a simple current system. *J. Meteorol.* **2(2)**, 113–119.

Charney, J. G. (1947) The dynamics of long waves in a baroclinic westerly current. *J. Meteorol.* **4(5)**, 135–162.

Charney, J. G. (1948) On the scale of atmospheric motions. *Geofys. Publikasjoner, Oslo* **17(2)**, 1–17.

Charney, J. G. (1955) The use of the primitive equations in numerical prediction. *Tellus* **7(1)**, 22–26.

Charney, J. G. (1971) Geostrophic turbulence. *J. Atmos. Sci.* **28(6)**, 1087–1095.

Charney, J. G. and A. Eliassen (1949) A numerical method for predicting the perturbations of the middle latitude westerlies. *Tellus* **1(2)**, 38–54.

Charney, J. G., R. Fjortoft and J. von Neumann (1950) Numerical integration of the barotropic vorticity equation. *Tellus* **2**, 237–254.

Charney, J. G. and M. E. Stern (1962) On the stability of internal baroclinic jets in a rotating atmosphere. *J. Atmos. Sci.* **19(2)**, 159–172.

Charney, J. G. and J. G. Devore (1979) Multiple flow equilibria in the atmosphere and blocking. *J. Atmos. Sci.* **36**, 1205–1216.

Chetaev, N. G. (1990) *Ustoychvist' Dvizheniya (Stability of Motion).* 4th ed. Nauka Publishers, Moscow, 176 pp (in Russian).

Courant, R. and D. Hilbert (1953) *Methods of Mathematical Physics.* Vol. 1, New York: Interscience Publ.

Craig, R. A. (1945) A solution of the nonlinear vorticity equation for atmospheric motion. *J. Meteorol.* **2(4)**, 175–178.

Crighton, D. G. (1981) Acoustics as a branch of fluid mechanics. *J. Fluid Mech.* **106**, 261–298.

Cushman-Roisin, B. (1994) *Introduction to Geophysical Fluid Dynamics.* Prentice Hall, London, 320 pp.

Danielsen, E. F. (1968) Stratospheric–tropospheric exchange based on radioactivity, ozone and potential vorticity. *J. Atmos. Sci.* **25**, 502–518.

Danielsen, E. F. (1990) In defense of Ertel's potential vorticity and its general applicability as a meteorological tracer. *J. Atmos. Sci.* **47**, 2013–2020.

Danilov, S. D. (1992) Turbulent Ekman layer effect on the stability of global atmospheric flows. *Izvestiya, Atmospheric Oceanic Physics* **28(12)**, 835–841 (English translation).

Diky, L. A. (1965) A nonlinear theory of hydrodynamic stability. *Prikl. Matematika i Mekhanika.* **29(5)**, 852–855 (in Russian).

Diky, L. A. (1969) A variational principle in the theory of meteorological-field adaptation. *Izvestiya, Atmospheric Oceanic Physics* **5(2)**, 98–100 (English translation).

Diky, L. A. (1972) A comment on the adiabatic-flow invariants suggested by Hollmann. *Izvestiya, Atmospheric Oceanic Physics* **8(3)**, 182–183 (English translation).

Diky, L. A. (1976) *Gidrodimamicheskaya Ustoychivost' i Dinamika Atmosferi (Hydrodynamic Stability and Dynamics of the Atmosphere).* Leningrad, Gidrometeoizdat, 108 pp (in Russian).

Diky, L. A. and M. V. Kurgansky (1971) Integral conservation law for perturbations of zonal flow, and its applications to stability studies. *Izvestiya, Atmospheric Oceanic Physics* **7(9)**, 623–626 (English translation).

Dobryshman, E. M., N. S. Filippova, M. I. Fortus and Ya. M. Kheifetz (1982) *Krupnomasshtabnye Statististicheskie Kharakteristiki Global'nogo Polya Prizemnogo Davleniya (Large-Scale Statistical Characteristics of the Surface Pressure Global Field).* Gidrometeoizdat Press, Moscow, 48 pp (in Russian).

Dolzhanskii, F. V. and G. S. Golitsyn (1977) Laboratory modeling of global geophysical flows: a review. *Izvestiya, Atmospheric Oceanic Physics* **13(8)**, 550–564 (English translation).

Dolzhanskii, F. V., V. A. Krymov and D. Yu. Manin (1990) Stability and vortex structures of quasi two-dimensional shear flows. *Usp. Fiz. Nauk* **160** (7), 1–47 (transl. in *Uspekhi Sov. Phys.* **33(7)**, 495–520).

Dolzhanskii, F. V., V. A. Krymov and D. Yu. Manin (1992) An advanced experimental investigation of quasi-two-dimensional shear flows. *J. Fluid Mech.* **241**, 705–722.

Dolzhanskii, F. V., M. V. Kurgansky and Yu. L. Chernous'ko (1979) Laboratory and theoretical study of barotropic Rossby waves in a rotating annulus. *Izvestiya, Atmospheric Oceanic Physics* **15(6)**, 408–414 (English translation).

Dolzhanskii, F. V. and D. Yu. Manin (1993) On the effect of turbulent Ekman layer on global atmospheric dynamics. *Geophys. Astrophys. Fluid Dyn.* **72**, 93–105.

Drazin, P. G. and W. H. Reid (1982) *Hydrodynamic Stability.* Cambridge University Press, 527 pp.

Dubuc, A. F. (1947) Towards vertical velocities computation from a pressure field. *Trudy Nauchno-Issledovatel'skikh Uchrezhdenii Glavnogo Upravleniya Gidrometeosluzhby.* Ser. II, Issue 24 (in Russian).

Dutton, J. A. (1973) The global thermodynamics of atmospheric motion. *Tellus* **25(2)**, 89–111.

Dutton, J. A. and D. R. Johnson (1967) The theory of available potential energy and a variational approach to atmospheric energetics. *Adv. Geophys.* **12**, 333.

Dymnikov, V. P. and A. N. Filatov (1990) *Ustoichivost' krupnomasshtabnykh atmosfernykh protsessov (Stability of large-scale atmospheric processes).* Leningrad, Gidrometeoizdat, 237 pp (in Russian).

Dymnikov, V. P. and A. N. Filatov (1997) *Mathematics of Climate Modeling.* Birkhaeuser, Boston–Basel–Berlin, 264 pp.

Eady, E. T. (1949) Long waves and cyclone waves. *Tellus* **1(3)**, 33–52.

Egger, J. (1989) A note on complete sets of material conservation laws. *J. Fluid Mech.* **204**, 543–548.

Egger, J. and K.-P. Hoinka (1999) The equatorial bulge, angular momentum and planetary wave motion. *Tellus* **51A**, 914–921.

Ekman, V. W. (1905) On the influence of the Earth's rotation on ocean-currents. *Ark. Mat. Astronom. Fys.* **2(11)**, 1–52.

Eliassen, A. (1948) The quasi-static equations of motion with pressure as independent variable. *Geofys. Publ.* **17(3)**, 44 pp.

Eliassen, A. (1987) Entropy coordinates in atmospheric dynamics. *Z. Meteorol.* **37(1)**, 1–11.

Eliassen, A. and E. Kleinschmidt (1957) Dynamic Meteorology. *Handbuch der Physik* **48**, 1–154.

Encyclopedia of Climate and Weather (1996) Ed. S. H. Schneider. Oxford University Press, Vol. 1 A–K, pp. 1–459, Vol. 2 L–Z, pp. 461–929.

Ertel, H. (1941) Über neue atmosphärische Bewegungsgleichungen und eine Differentialgleichung des Luftdruckfeldes (On a new equation of atmospheric motion and a differential equation of the air pressure field). *Meteorol. Z.* **58(3)**, 77–78. [See also *Theoretical Meteorology, Weather Prediction, Cosmology and General Applications.* Eds. W. Schröder and H.-J. Treder, Bremen–Rönnebeck, 156 pp., 1995.]

Ertel, H. (1942) Ein neuer hydrodynamischer Wirbelsatz (A new hydrodynamic vorticity equation). *Meteorol. Z.* **59(9)**, 277–281. [See also: *Geophysical Hydrodynamics and Ertel's Potential Vorticity (Selected papers of Hans Ertel),* Ed. W. Schröder, Bremen–Rönnebeck, 218 pp., 1991.]

Ertel, H. (1943) Über stationäre oszillatorische Luftströmungen auf der rotierden Erde (About stationary oscillating air flows on the rotating Earth). *Meteorol. Z.* **60**, 332–334.

Ertel, H. (1960) Relación entre la derivada individual y una cierta divergencia especial en Hidrodinámica (Relation between the individual derivative and a certain spatial divergence in hydrodynamics). *Gerlands Beiträge zur Geophysik* **69**, 290–293. [See also: *Geophysical Hydrodynamics and Ertel's Potential Vorticity (Selected papers of Hans Ertel).* Ed. W. Schröder, Bremen–Rönnebeck, 218 pp, 1991.]

Ertel, H. and C.-G. Rossby (1949) A new conservation theorem of hydrodynamics. *Geofisica pura e applicata–Milano.* **14**, 189–193.

Etling, D. (1985) Some aspects of helicity in atmospheric flows. *Beitr. Phys. Atmosph.* **58(1)**, 88–100.

Etling, D. (1996) *Theoretische Meteorologie. Eine Einfürung (Theoretical Meteorology. An Introduction).* Vieweg and Sohn, Braunschweig/Wiesbaden. 318 pp.

Fjortoft, R. (1950) Application of integral theorems in deriving criteria of stability of laminar flows and of baroclinic circular vortex. *Geofys. Publikasjoner* **17(5)**, 1–52.

Ford, R., M. E. McIntyre and W. Norton (2000) Balance and the slow manifold: Some explicit results. *J. Atmos. Sci.* **57**, 1236–1254.

Friedmann, A. A. (1934) *Opyt Gidromekhaniki Szhimaemoi Zhidkosti (Essay on Hydromechanics of Compressible Fluid).* ONTI–GTTI Press, Leningrad–Moscow, 370 pp (in Russian).

Frisch, U. (1998) *Turbulence: the Legacy of A. N. Kolmogorov.* Cambridge University Press, 296 pp.

Galin, M. B. and S. E. Kirichkov (1985) Stability of atmospheric zonal circulation in a model including orography and the blocking problem. *Izvestiya, Atmospheric Oceanic Physics* **21(6)**, 433–439 (English translation).

Gibbs, J. W. (1873) A method of geometrical representation of the thermodynamic properties of substance by means of surfaces. *Trans. Connecticut Acad.* **2**, 382–404. (pp. 33–54 in Vol. 1 of *The Collected Works of J. W. Gibbs*, 1928. Longmans Green and Co.)

Gill, A. E. (1982) *Atmosphere–Ocean Dynamics.* Academic Press, 662 pp.

Gledzer, Ye. B., F. V. Dolzhaskii and A. M. Obukhov (1981) *Sistemy Gidrodinamicheskogo Tipa i ikh Primennie (Hydrodynamic-Type Systems and Their Applications).* Nauka Publishers, Moscow, 368 pp (in Russian).

Goldstein, H. (1980) *Classical Mechanics.* Addison–Wesley Press, Cambridge, MA, 672 pp.

Golitsyn, G. S. (1973) *Vvedenie v Dinamiku Planetnykh Atmosfer (An Introduction to the Dynamics of the Planetary Atmospheres).* Gidrometeoizdat, Leningrad, 104 pp (in Russian).

Griffiths, R. W. and E. J. Hopfinger (1986) Experiments with baroclinic vortex pairs in a rotating fluid. *J. Fluid Mech.* **173**, 501–518.

Gryanik, V. M. (1983) Emission of sound by linear vortical filaments. *Izvestiya, Atmospheric Oceanic Physics* **19(2)**, 150–152 (English translation).

Gryanik, V. M. (1983a) Dynamics of singular geostrophic vortices in a two-layer model of the atmosphere (Ocean). *Izvestiya, Atmospheric Oceanic Physics* **19(3)**, 171–179 (English translation).

Gryanik, V. M. (1983b) Dynamics of localized vortex perturbations "vortex charges" in a baroclinic fluid. *Izvestiya, Atmospheric Oceanic Physics* **19(5)**, 347–352 (English translation).

ter Haar, D. (1964) *Elements of Hamiltonian Mechanics.* 2nd ed., Pergamon Press.

Haltiner, G. J. and F. L. Martin (1957) *Dynamic and Physical Meteorology.* McGraw Hill Book Co., 470 pp.

Haltiner, G. J. and R. T. Williams (1980) *Numerical Prediction and Dynamic Meteorology.* 2nd ed., John Wiley and Sons, 477 pp.

Hartmann, D. L. (1995) A PV view of zonal flow vacillation. *J. Atmos. Sci.* **52**, 2561–2576.

Haurwitz, B. (1940) The motion of atmospheric disturbances on a spherical earth. *J. Marine Res.* **3**, 254–267.

Haynes, P. H. and M. E. McIntyre (1987) On the evolution of vorticity and potential vorticity in the presence of diabatic heating and frictional or other forces. *J. Atmos. Sci.* **44(5)**, 828–841.

Haynes, P. H. and M. E. McIntyre (1990) On the conservation and impermeability theorems for potential vorticity. *J. Atmos. Sci.* **47(16)**, 2021–2031.

Hess, P. Q. (1991) Mixing processes following the final atmospheric warming. *J. Atmos. Sci.* **48(14)**, 1625–1641.

Hide, R. (1989) Superhelicity, helicity and potential vorticity. *Geophys. Astrophys. Fluid Dyn.* **48**, 69–79.

Hide, R. and P. J. Mason (1975) Sloping convection in a rotating fluid. *Adv. Phys.* **24(1)**, 47–100.

Hide, R., N. T. Birch, L. V. Morrison *et al.* (1980) Atmospheric angular momentum fluctuations and changes in the length of day. *Nature* **286**, 114–117.

Hollmann, G. H. (1964) Ein vollständiges System hydrodynamischer Erhaltungssätze (A complete system of hydrodynamic conservation laws). *Archiv Meteorol. Geophys. Bioklimatol.* **A14(1)**, 1–13.

Holloway, G. (1993) The role of oceans in climate change: a challenge to large eddy simulation. In: *Large Eddy Simulation of Complex Engineering and Geophysical Flows*. Eds. B. Galperin and S. A. Orszag. Cambridge University Press, 425–440.

Holton, J. R. (1992) *An Introduction to Dynamic Meteorology*. 3rd ed., Academic Press (San Diego), 507 pp.

Holton, J. R., P. H. Haynes, M. E. McIntyre, A. R. Douglass, R. B. Rood and L. Pfister (1995) Stratosphere–troposphere exchange. *Rev. Geophys.* **33(4)**, 403–439.

Hoskins, B. (1975) The geostrophic momentum approximation and the semi-geostrophic equations. *J. Atmos. Sci.* **32**, 233–242.

Hoskins, B. J. (1991) Towards a PV-theta view of the general circulation. *Tellus, Ser. A* **43**, 27–35.

Hoskins, B. J., M. E. McIntyre and A. W. Robertson (1985) On the use and significance of isentropic potential vorticity maps. *Quart. J. Roy. Meteorol. Soc.* **111**, 877–946.

Hough, S. S. (1898) On the application of harmonic analysis to the dynamical theory of the tides, II. On the general integration of Laplace's tidal equations. *Phil. Trans. Roy. Soc. Lond.* A **191**, 139–185.

Howard, L. N. (1961) Note on a paper of John W. Miles. *J. Fluid Mech.* **10**, Pt. 4, 509–512.

Ingel', L. Kh. and L. A. Mikhailova (1990) Theory of the Ekman boundary layer with non-linear boundary conditions. *Izvestiya, Atmospheric Oceanic Physics* **26(7)**, 499–503 (English translation).

James, I. N. (1994) *Introduction to Circulating Atmospheres*. Cambridge University Press, 422 pp.

Juckes, M. N. and M. E. McIntyre (1987) A high-resolution one-layer model of breaking planetary waves in the stratosphere. *Nature* **328(6131)**, 590–596.

Källen, E. (1984) Bifurcation mechanisms and atmospheric blocking. *Problems and prospects in long and medium range weather forecasting. Eds. D. Burridge and E. Källen.* P. 229–263.

Kibel', I. A. (1957) *Vvedenie v Gidrodinamicheskie Metody Kratkosrochnogo Prognoza Pogody (Introduction to Hydrodynamic Methods for Short-Term Weather Prediction)*. Moscow, Gostekhizdat, 376 pp (in Russian).

Klyatskin, V. I. (1966) Sound emmision by a system of vortices. *Izv. Akad. Nauk SSSR Mekhanika zhidkosti i gaza* **6**, 87–92 (in Russian).

Kochin, N. E., I. A. Kibel' and N. V. Rose (1964) *Theoretical Hydromechanics,* Interscience Publishers, 577 pp.

Koschmieder, H. (1933) *Dynamische Meteorologie (Dynamic Meteorology).* Leipzig.

Kuo, H. L. (1949) Dynamic instability of two-dimensional nondivergent flow in a barotropic atmosphere. *J. Meteorol.* **6(2)**, 105–122.

Kurgansky, M. V. (1981) On the integrated energy characteristics of the atmosphere. *Izvestiya, Atmospheric Oceanic Physics* **17(9)**, 686–692 (English translation).

Kurgansky, M. V. (1983) On the synthesis of the static stability integral characteristics of the atmosphere. *Izvestiya, Atmospheric Oceanic Physics* **19(7)**, 570–573 (English translation).

Kurgansky, M. V. (1988) Shear flow instability in a stratified fluid. *Izvestiya, Atmospheric Oceanic Physics* **24(5)**, 343–348 (English translation).

Kurgansky, M. V. (1989) Relationship between helicity and potential vorticity in a compressible rotating fluid. *Izvestiya, Atmospheric Oceanic Physics* **25(12)**, 979–981 (English translation).

Kurgansky, M. V. (1991) The atmospheric vortex charge. *Izvestiya, Atmospheric Oceanic Physics* **27(7)**, 510–516 (English translation).

Kurgansky, M. V. and D. S. Prikazchikov (1994) Potential vorticity as an indicator of seasonal variability of the atmosphere. *Izvestiya, Atmospheric Oceanic Physics* **30(6)**, 696–703.

Kurgansky, M. V. and I. A. Pisnichenko (1997) Potential vorticity as an atmospheric climate characteristic. *Doklady Akad. Nauk* **357(1)**, 104–107.

Kurgansky, M. V. and I. A. Pisnichenko (2000) Modified Ertel potential vorticity as a climate variable. *J. Atmos. Sci.* **57**, 822–835.

Kurgansky, M. V. and M. S. Tatarskaya (1987) The potential vorticity concept in meteorology: a review. *Izvestiya, Atmospheric Oceanic Physics* **23(8)**, 587–606 (English translation).

Kurgansky, M. V. and M. S. Tatarskaya (1990) Adiabatic invariants of large-scale atmospheric processes. *Izvestiya, Atmospheric Oceanic Physics* **26(12)**, 894–901.

Kurgansky, M. V. and M. S. Tatarskaya (1990a) Diagnosis of atmospheric circulation on the basis of FGGE data with the help of potential vorticity. In:*Itogi Nauki I Tekhniki VINITI. Ser.: Atmosfera, Okean, Kosmos – programma 'Razrezy',* **13**, 4–12. (in Russian).

Kurgansky, M. V., K. Dethloff, I. A. Pisnichenko, H. Gernandt, F.-M. Chmielewski and W. Jansen (1996) Long-term climate variability in a simple, nonlinear atmospheric model. *J. Geophys. Res.* **101(D2)**, 4299–4314.

Lait, L. R. (1994) An alternative form for potential vorticity. *J. Atmos. Sci.* **51**, 1754–1759.

Lamb, H. (1945) *Hydrodynamics.* 6th ed., Dover, New York, 738 pp.

Landau, L. D. and E. M. Lifshitz (1973) *Mekhanika (Mechanics).* Nauka Publishers, Moscow, 208 pp (in Russian). Published in English as *Mechanics.* Addison–Wesley, Reading, Mass. 1960.

Landau, L. D. and E. M. Lifshitz (1988) *Gidromekhanika (Hydromechanics).* Nauka Publishers, Moscow, 736 pp (in Russian). Published in English as *Fluid Mechanics.* 2nd Ed. Pergamon Press, Oxford, 1987. Addison–Wesley, Reading, Mass. 1960.

Larichev, V. D. and A. B. Fedotov (1988) Phenomenon of self-organization in 2D-turbulence on a beta-plane. *Doklady Akad. Nauk SSSR* **298(4)**, 971–975 (in Russian).

Leech, J. W. (1958) *Classical Mechanics*. Methuen's Monographs on Physical Subjects. London: Methuen and Co LTD, NY: Wiley and Sons Inc.

Lesieur, M. (1997) *Turbulence in Fluids*. 3rd ed., Kluwer Academic Publishers, 515 pp.

Lilly, D. K. (1982) The development and maintenance of rotation in convective storms. In: *Topics in Atmospheric and Oceanographic Sciences. Intense Atmospheric Vortices*. Eds L. Bengtsson and J. Lighthill. Springer–Verlag. Berlin–Heidelberg, pp. 149–160.

Lilly, D. K. (1985) The structure, energetics and propagation of rotating convective storms. Part II: Helicity and storm stabilization. *J. Atmos. Sci.* **42(2)**, 126–140.

Lin, C. C. (1966) *The Theory of Hydrodynamic Stability*. Cambridge University Press, 155 pp.

Lindzen, R. (1990) *Dynamics in Atmospheric Physics*. Cambridge University Press, Cambridge, 310 pp.

Loitsyansky, L. G. (1973) *Mekhanika Zhidkosti i Gaza (Mechanics of Liquids and Gases)*. Nauka Publishers, Moscow, 848 pp (in Russian).

Longuet-Higgins, M. S. (1968) The eigenfuctions of Laplace's tidal equations over a sphere. *Phil. Trans. Roy. Soc. Lond.* A **262**, 511–607.

Lorenz, E. N. (1955) Available potential energy and the maintenance of the general circulation. *Tellus* **7**, 157–167.

Lorenz, E. N. (1960) Maximum simplification of the dynamical equations. *Tellus* **12(3)**, 243–254.

Lorenz, E. N. (1967) *The Nature and Theory of the General Circulation of the Atmosphere*. World Meteorological Organization, Geneva, 161 pp.

Lorenz, E. N. (1995) *The Essence of Chaos*. University of Washington Press, 227 pp.

Lynch, P. (1989) The slow equations. *Quart. J. Roy. Meteorol. Soc.* **115**, 201–219.

Magnusdottir, G. and P. H. Haynes (1996) Wave activity diagnostics applied to baroclinic wave life cycles. *J. Atmos. Sci.* **53**, 2317–2353.

Marquet, P. (1991) On the concept of exergy and available enthalpy: Application to atmospheric energetics. *Quart. J. Roy. Meteorol. Soc.* **117(499)**, 449–475.

McIntyre, M. E. (2001) Balance, potential-vorticity inversion, Lighthill radiation, and the slow quasimanifold. In: *Advances in Mathematical Modeling of Atmosphere and Ocean Dynamics* (Proc. IUTAM Limerick Symposium), ed. P. F. Hodnett. Dordrecht, Kluwer Publishers, 45–68.

McIntyre, M. E. and W. A. Norton (1989) Potential vorticity inversion on a hemisphere. 5th Sci. Assembly IAMAP. Univ. Reading, UK. 31 July–12 August 1989. Brief rev. papers and abstracts.

McIntyre, M. E. and W. A. Norton (1990) Dissipative wave-mean interactions and the transport of vorticity or potential vorticity. *J. Fluid Mech.* **212**, 403–435.

McIntyre, M. E. and W. Norton (2000) Potential vorticity inversion on a hemisphere. *J. Atmos. Sci.* **57**, 1214–1235.

McIntyre, M. E. and T. N. Palmer (1983) Breaking planetary waves in the stratosphere. *Nature* **305**, 593–600.

McIntyre, M. E. and T. N. Palmer (1984) The "surf zone" in the stratosphere. *J. Atmos. Terr. Phys.* **46(9)**, 825–849.

Miles, J. W. (1961) On the stability of heterogeneous shear flows. *J. Fluid. Mech.* **10**, Pt. 4, 496–509.

Miles, J. W. (1986) Richardson's criterion for the stability of stratified shear flow. *Phys. Fluids.* **29(10)**, 3470–3471.

Milne-Thompson, L. M. (1960) *Theoretical Hydromechanics.* 4th ed., London, Macmillan and Co. LTD New York.

Moffatt, H. K. (1969) The degree of knottedness of tangled vortex lines. *J. Fluid Mech.* **35**, Pt. 1, 117–129.

Monin, A. S. (1972) *Weather Prediction as a Problem of Physics.* MIT Press Cambridge, MA, 199 pp.

Monin, A. S. (1990) *Theoretical Geophysical Fluid Dynamics.* Dordrecht Kluwer, 399 pp.

Monin, A. S. and A. M. Obukhov (1954) Basic laws of turbulent mixing in the ground layer of the atmosphere. *Akad. Nauk SSSR Geofiz. Inst. Tr.* **151**, 163–187 (in Russian).

Monin, A. S. and A. M. Yaglom (1971, 1975) *Statistical Fluid Mechanics* 1+2. Ed. J. Lumley. The MIT Press, Cambridge, Massachusetts (769 + 847 pp).

Montgomery, R. B. (1939) A suggested method for representing gradient flow in isentropic surfaces. *Bull. Amer. Meteorol. Soc.* **18**, 210–212.

Moran, F. (1942) Nueva deduccion e interpretation fisica de un teorema hidrodinamico de Ertel (New derivation and physical interpretation of Ertel's hydrodynamic theorem). *Revista di Geofisica (Madrid),* 3–7. [see *Theoretical Concepts and Observational Implications in Meteorology and Geophysics.* Eds. W. Schroeder and H.-J. Treder, Bremen–Rönnebeck, 206 pp, 1993.]

Moreau, J. J. (1961) Constantes d'un ilot tourbillonnaire en fluide parfait barotrope (Constants of a turbulent flow of ideal barotropic fluid). *C. R. Acad. Sci. Paris* **252**, 2810–2812.

Müller, P. (1995) Ertel's potential vorticity theorem in physical oceanography. *Rev. Geophys.* **33(1)**, 67–97.

Nakamura, N. (1995) Modified Lagrangian-mean diagnostics of the stratospheric polar vortices. Part I: Formulation and analysis of GFDL SKYHI GCM. *J. Atmos. Sci.* **52**, 2096–2108.

Namias, J. (1939) The use of isentropic analysis in short-term forecasting. *J. Aeronautical Sci.* **6**, 295–298.

Neamtan, S. M. (1946) The motion of harmonic waves in the atmosphere. *J. Meteorol.* **3(2)**, 43–56.

Nezlin, M. V. (1986) Rossby solitons. *Usp. Fiz. Nauk* **150 (1)**, 3–60 (in Russian).

Obukhov, A. M. (1946) Turbulence in the thermally inhomogeneous atmosphere. *Trudy Inst. Teor. Geofiziki* **1**, 95–115 (in Russian).

Obukhov, A. M. (1949) Structure of the temperature field in a turbulent flow. *Izv. Akad. Nauk SSSR. Ser. Geograf. Geofiz.* **13(1)**, 58–69 (in Russian).

Obukhov, A. M. (1949a) On the question of geostrophic wind. *Bull. USSR Acad. Sci. Geograph. Geophys. Series* **13(4)**, 281–306 (in Russian).

Obukhov, A. M. (1962) On the dynamics of a stratified fluid. *Dokl. Akad. Nauk SSSR* **145(6)**, 1239–1242 (in Russian).

Obukhov, A. M. (1964) Adiabatic invariants for atmospheric processes. *Meteorologiya i Gidrologiya* **2**, 3–9 (in Russian).

Obukhov, A. M. (1984) On potential vorticity. In: *N. E. Kochin and the Development of Mechanics* (in Russian). Ed. A. Yu. Ishlinslii. Nauka Publishers, Moscow, 84–93 (in Russian).

Obukhov, A. M. (1988) *Turbulentnost' i Dynamika Atmosfery (Turbulence and Dynamics of the Atmosphere)*. Leningrad, Gidrometeoizdat, 414 pp (in Russian).

Obukhov, A. M., M. V. Kurgansky and M. S. Tatarskaya (1988) Isentropic analysis of global atmospheric processes using the potential vorticity field from FGGE data. *Meteorologyia i Gydrologiya* **8**, 111–119 (in Russian).

Oort, A. H. (1964) On estimates of the atmospheric energy cycle. *Month. Wea. Rev.* **92**, 483–493.

Oort, A. H. (1983) Global atmospheric circulation statistics 1958–1973, *NOAA Professional Paper 14*, US Dept of Commerse, 180 pp + 47 microfishes.

Orlanski, I. (1975) A rational subdivision of scales for atmospheric processes. *Bull. Amer. Met. Soc.* **56**, 527–530.

Ozmidov, R. V. (1983) Small-scale turbulence and fine structure of hydrophysical fields in the ocean. *Oceanology* **23(4)**, 533–537 (in Russian).

Paegle, J. H. (1979) The effect of topography on Rossby wave. *J. Atmos. Sci.* **36**, 2267–2271.

Palmen, E. and C. W. Newton (1969) *Atmospheric Circulation Systems (Their Structure and Physical Interpretation)*. Academic Press, New York and London, 603 pp.

Pearce, R. P. (1978) On the concept of available potential energy. *Quart. J. Roy. Meteorol. Soc.* **104(441)**, 737–755.

Pedlosky, J. (1987) *Geophysical Fluid Dynamics*. 2nd ed., Springer-Verlag, New York, 710 pp.

Peixoto, J. P. and A. H. Oort (1992) *Physics of Climate*. American Institute of Physics, New York, 520 pp.

Petoukhov, V., A. Ganopolski, V. Brovkin, M. Claussen, C. Kubatzki and S. Rahmstorf (2000) CLIMBER-2: a climate model of intermediate complexity. Part I: Model description and performance for present climate. *Climate Dynamics* **16**, 1–7.

Phillips, N. A. (1951) A simple three-dimensional model for the study of large-scale flow patterns. *J. Meteorol.* **8**, 381–394.

Phillips, N. A. (1957) A coordinate system having some special advantage for numerical forecasting. *J. Meteorol.* **14(2)**, 184–185.

Phillips, N. A. (1963) Geostrophic motions. *Rev. Geophys.* **1(2)**, 123–176.

Pielke, R. A. (1984) *Mesoscale Meteorological Modeling*. Academic Press, 612 pp.

Pierrehumbert, R. (1991) Large-scale horizontal mixing in planetary atmosphere. *Phys. Fluids. Ser. A.* **3(5)** Pt. 2, 1250–1260.

Pisnichenko, I. A. (1986) Inclusion of orography in the problem of the motion of a barotropic atmosphere over a spherical Earth. *Izvestiya, Atmospheric Oceanic Physics* **22(10)**, 792–797 (English translation).

Pisnichenko, I. A. and M. V. Kurgansky (1996) Adiabatic invariants and diagnostic studies of climate. *An. Acad. Bras. Ci.* **68** (Suppl. 1), 261–277.

Rayleigh, Lord (1880) On the stability, or instability, of certain fluid motions. *Sci. Papers. Cambridge Univ. Press* **1**, 474–487.

Rayleigh, Lord (1916) On convection currents in a horizontal layer of fluid when the higher temperature is on the under side. *Sci. Papers. Cambridge Univ. Press.* **6**, 432–443.

Reed, R. J. (1955) A study of a characteristic type of upper-level frontogenesis. *J. Meteorol.* **12**, 226–237.

Reed, R. J. and E. F. Danielsen (1959) Fronts in the vicinity of the tropopause. *Arch. Meteorol. Geophys. Biokl.* **A11**, 1–17.

Rochas, M. (1986) A new class of exact time-dependent solutions of the vorticity equation. *Month. Wea. Rev.* **114**, 961–966.

Rossby, C.-G. (1938) On the mutual adjustment of pressure and velocity distributions in certain simple current systems, II. *J. Marine Res.* **1(3)**, 239–263.

Rossby, C.-G. (1940) Planetary flow patterns in the atmosphere. *Quart. J. Roy. Meteorol. Soc.* **66** Suppl., 68–87.

Rossby, C.-G., D. P. Keily, J. W. W. Osmun and J. Namias (1937) Isentropic analysis. *Bull. Amer. Meteorol. Soc.* **18**, 201–209.

Rossby, C.-G. and collaborators (1939) Relation between variations in the intensity of the zonal circulation of the atmosphere and the displacements of the semi-permanent centers of action. *J. Marine Res.* **2(1)**, 38–55.

Rotunno, R. and M. Fontini (1989) Petersen's 'Type B' cyclogenesis in terms of discrete neutral Eady modes. *J. Atmos. Sci.* **46(23)**, 3599–3604.

Ryym, R. Y. (1990) General form of equation of atmospheric dynamics in isobaric coordinates. *Izvestiya, Atmospheric Oceanic Physics* **26(1)**, 9–14 (English translation).

Salmon, R. (1998) *Lectures on Geophysical Fluid Dynamics.* Oxford University Press, 378 pp.

Schmidt, W. (1925) *Der Massenaustauch in freier Luft und verwandte Erscheinungen (Mass Exchange in the Free Atmosphere and Similar Phenomena).* Hamburg.

Schröder, W. (1988) An additional bibliographical sketch on the development of Ertel's potential vorticity theorem. *Quart. J. Roy. Meteorol. Soc.* **114**(484), 1563–1567.

Shakina, N. P. (1990) *Gidrodinamicheskaya Neustoichivost' v Atmosfere (Hydrodynamic Instability in the Atmosphere).* Leningrad, Gidrometeoizdat, 308 pp (in Russian).

Sidorenkov, N. S. (1976) A study of the angular momentum of the atmosphere. *Izvestiya, Atmospheric Oceanic Physics* **12(6)**, 351–356 (English translation).

Slezkin, N. A. (1990) Hydrodynamic model for a typhoon taking into account the Earth's rotation. *Izv. Akad. Nauk SSSR, Fizika atmosphery i okeana* **26**(5), 364–369 (English translation).

Sretensky, L. N. (1987) *Dinamicheskaya Teoriya Prilivov (Dynamic Theory of Tides).* Nauka Publishers, Moscow, 472 pp (in Russian).

Starr, V. (1966) *Physics of Negative Viscosity Phenomena.* Mc Graw-Hill, New York, 256 pp.

Stone, P. H. (1978) Baroclinic adjustment. *J. Atmos. Sci.* **35**(4), 561–571.

Tatarskaya, M. S. (1978) On the adiabatic invariant distribution in the atmosphere of the Northern Hemisphere. *Izvestiya, Atmospheric Oceanic Physics* **15**, 67–71 (English translation).

Taylor, G. I. (1914) Eddy motion in the atmosphere. *Phil. Trans. Roy. Soc. London* **A215**, 1–26 (Scientific papers 2, *1*).

Tennekes, H. (1977) The general circulation of two-dimensional turbulent flow on a beta-plane. *J. Atmos. Sci.* **34**, 702–712.

Thompson, P. D. (1961) *Numerical Weather Analysis and Prediction.* Macmillan, New York, 170 pp.

Thompson, P. D. (1982) A generalized class of time-dependent solutions of the vorticity equation for nondivergent barotropic flow. *Mon. Wea. Rev.* **110**, 1321–1324.

Turner, J. S. (1973) *Buoyancy Effects in Fluids.* Cambridge Univ. Press, 368 pp.

Van Mieghem, J. (1973) *Amospheric Energetics.* Oxford Monographs on Meteotology, Clarendon Press, Oxford, 306 pp.

Verkley, W. T. M. (1984) The construction of barotropic modons on a sphere. *J. Atmos. Sci.* **41**, 2492–2504.

Wallace, J. M. and P. V. Hobbs (1977) *Atmospheric Science. An Introductionary Survey.* Academic Press, Inc., 452 pp.

Wiin-Nielsen, A. and T.-C. Chen (1993) *Fundamentals of Atmospheric Energetics.* Oxford Univ. Press, 372 pp.

Williams, G. P. (1974) Generalized Eady waves. *J. Fluid Mech.* **62**, Pt. 4, 643–655.

Yudin, M. I. (1955) Invariant quantities in large-scale atmospheric processes. *Trudy Glavnoi Geofizicheskoi Observatorii* **55(117)**, 3–12 (in Russian).

Zeytounian, R. K. (1991) *Meteorological Fluid Dynamics (Asymptotic Modelling, Stability and Chaotic Atmospheric Motion).* Lecture Notes in Physics, New Series: Monographs, Vol. 5. Springer-Verlag Berlin Heidelberg, 346 pp.

INDEX

adiabatic approximation, 5, 19, 106,
126, 130, 134, 140, 141
adiabatic invariants, 5, 19, 20, 89, 141,
154, 167
in classical mechanics, 6
anticyclogenesis (diabatic), 137, 139
anticyclones, 2, 4, 174, 199, 200;
see also atmospheric blocking
Arnol'd method of stability analysis, 89
atmospheric angular momentum, *see*
conservation laws
atmospheric blocking, 155, 158, 159,
163, 166, 174, 181
atmospheric free energy, 35, 37, 38, 42,
55
atmospheric heat engine, 31, 32, 38, 49,
185
attractor of climate system, 11, 181,
182
available enthalpy, 55
available kinetic energy, 50, 116, 120
available potential energy, 34, 35, 38,
98, 116, 123, 124, 133, 134, 135,
142

baroclinic instability, 96, 97, 115, 116
in Eady's model, 106–110
in Phillip's model, 113, 114
in two-dimensional model, 57
barotropic instability, 89, 96, 115
Beltrami flow, 31
Bernoulli function, 15, 129
beta-effect, 72, 75, 90, 110, 205
beta-plane approximation, 97

Bjerknes' circulation theorem, 57, 151
boundary layer, 45, 46
surface layer, 187–193
Ekman layer, 172, 173,
193–199, 201
Boussinesq approximation, 17, 191
Brunt–Vaisala frequency, 3, 4, 70, 116,
117, 193
Buys Ballot law, 63; *see also*
geostrophic wind

Carnot cycle, 35
Casimir functionals, 94
Charney–Obukhov's equation, 75,
201, 205, 206
Charney–Stern criterion, 104, 118; *see*
also baroclinic instability
conservation laws,
of absolute vorticity, 79, 93, 96,
129, 147
of angular momentum, 5, 43–50,
61, 90, 93, 96, 97, 176, 177
of energy, 5, 39, 54, 66–68, 85,
87, 90, 98, 108, 112, 130
of helicity, 26, 27, 29–31,
195–197, 205
of momentum, 85, 94, 99, 105,
112, 120
of potential vorticity, 5, 22, 23,
70, 78, 82, 96
of potential vorticity in
quasi-geostrophic
approximation, 75, 97, 110;
see also quasi-geostrophic

potential vorticity
  of potential vorticity in the
    shallow water model, 60, 61,
    64, 84
  of vorticity charge, 23–26, 61,
    129, 169–172, 174, 175, 182
Clebsch transformation of fluid
  dynamic equations, 153
Coriolis parameter, 2, 4, 52, 60, 63,
  72–74, 78, 105, 107, 205
cyclones, 2, 4, 155, 157, 159, 163, 181,
  199, 200
cyclogenesis (diabatic), 137, 139

deformation radius,
  Rossby's internal (baroclinic), 3,
    4, 70, 77
  Rossby–Obukhov's external
    (barotropic), see Obukhov's
    synoptic scale

Ekman boundary layer, see boundary
  layer
Ekman number, 201
Ekman pumping (suction), 199, 200
Ekman spiral, 195, 197, 199, 200
energy, see conservation laws
enstrophy, 90, 180
enthalpy, 12, 32
entropy, 12, 14, 20, 35, 38, 56
  informational, 169, 176, 177,
    180
entropy deficit, 40, 43
  information theoretic, 177, 182
Ertel's commutation formula, 153
Ertel's potential vorticity, 20, 21, 25,
  30, 31, 61, 128, 136, 141, 142–144,
  153, 170; see also conservation laws
  and potential vorticity
Euler's equations for a rigid rotator, 89
Exner function, 14, 125, 126

Friedmann's theorem, 17, 145, 152,
  178, 184
Friedmann's vorticity equation, 15, 22,
  136

Galilean invariance (principle of), 63,
  190, 194
general atmospheric circulation, 2, 43
geostrophic adjustment, 62, 63, 67, 68,
  70–72, 82
geostrophic balance, 62, 65, 66, 68, 72,
  73, 116
geostrophic singular (point) vortex, 70
geostrophic wind, 62, 63, 173, 195,
  198, 202–204
Gibbs' potential (free enthalpy), 12
Gromeka–Lamb equations, 15, 129

Headly's regime of atmospheric
  general circulation, 2
helicity, see conservation laws
Helmholtz' operator (Helmholtzian),
  16, 145
Helmholtz' vorticity equation, 80, 81
Hollmann's invariants, 30
horizontal baroclinicity, 57
Hough functions, 81, 85
hurricanes (typhoons), 2
hydrodynamic instability, 50, 89
  baroclinic, see baroclinic
    instability
  barotropic, see barotropic
    instability
  orographic, see orographic
    instability
  shear instability of stratified
    flows, see Richardson's
    criterion of stability
hyperviscosity, 11

inertial oscillations, 66
internal energy, 11, 32
  of a two-dimensional
    atmosphere, 56
isentropic analysis, 5, 123, 143, 149
isentropic coordinates, 35, 124, 127,
  128, 130, 132, 142, 151, 154
isentropic potential vorticity maps, 145,
  154, 155
invariant flux tube, 149, 151–155,
  158–167

jet stream, 1, 163

Kelvin's circulation theorem, 3, 21, 142, 147
Kibel'–Rossby parameter (number), 74, 77, 78
Kolmogorov's microscale of turbulence, 1
Kolmogorov–Obukhov's spectral law, 202

laboratory experiments, 62, 70, 110
Lagrangian invariants, 20, 138, 153, 155
Laplace's tidal operator, 81
length-of-day fluctuations, 46, 47, 177
Lighthill's theory of sound wave generation, 82
Lyapunov's method in the stability theory, 89, 92, 95, 96, 102, 104, 113, 118

Mach number, 78, 86
Margules theorem, 34, 37, 54
Miles–Howard criterion of stability, 118
Monin–Obukhov's length in the boundary layer theory, 193
Montgomery stream function, 126, 129

Newtonian (internal) viscosity, 10, 81, 201
Noether's theorem, 6

Obukhov's synoptic scale, 70, 71, 74, 87, 107, 111; see also deformation radius
orographic instability, 96
orographic torque, 45, 49

particle-relabelling symmetry, 6
planetary boundary layer, 3, 4, 136; see also boundary layer
planetary vorticity, 4, 104; see also Coriolis parameter
point vorticity charge, 26, 77; see also

conservation laws
potential enstrophy, 201
potential temperature, 14, 20, 56, 57, 61, 78, 98, 104, 121, 124, 125, 128, 143, 174, 191, 192
potential vorticity, see also conservation laws and Ertel's potential vorticity
  in isentropic coordinates, 124, 130, 137–141
  in isobaric coordinates, 128
  modified after Obukhov, 141, 142, 154–170, 174–185
Prandtl (logarithmic) boundary layer, 190
Prandtl number, 192
pressure (isobaric) coordinates, 75, 123, 127, 128
pseudo-potential vorticity, 5

quasi-biennial oscillation, 2, 181
quasi-geostrophic approximation, 17, 51, 55, 63, 78, 79, 97, 109, 115, 135, 171, 200
quasi-geostrophic potential vorticity, 5, 77, 200; see also conservation laws
quasi-solenoidal approximation, 78, 93, 96
quasi-static approximation, 17, 33, 52, 58, 75, 123, 124, 128, 133, 171

Rayleigh–Kuo's criterion, 95, 118; see also barotropic instability
Rayleigh (linear) friction, 11
Rayleigh's criterion of the onset of convection, 117
relative helicity, 31
Reynolds number, 187
Richardson number, 116, 119, 122, 128, 137
Richardson's criterion of stability, 116–118, 121, 191, 205
Rossby's deformation radius, see deformation radius
Rossby's regime of atmospheric general circulation, 2

Rossby waves, 51, 80, 81, 85, 97
    their stability, 95

semi-geostrophic equations, 79, 110
semi-permanent centres of action, 81
shallow-water model, 52, 59, 62, 82, 85
singular (point) vortices, 70, 82, 85
slow invariant manifold, 82
spontaneous wave emission, 51, 82, 86
super-rotation of the atmosphere, 46, 47, 49
surface friction torque, 45, 49

thermal wind relation, 3, 47, 53, 58, 107, 121, 128, 151, 176
topological invariants,
    of vorticity field, 29; *see also* helicity
    of potential vorticity field on isentropic surfaces, 145, 158, 159, 166
total enthalpy, 34, 38, 42, see also total potential energy
total potential energy, 34, 37, 38, 41, 56, 133, 135; *see also* Margules theorem

unavailable kinetic energy, 49
unavailable potential energy, 135

vorticity charge, *see* conservation laws

Printed and bound by CPI Group (UK) Ltd, Croydon, CR0 4YY

23/10/2024

01778238-0002